KNOWLEDGE DISCOVERY
AND MEASURES OF INTEREST

THE KLUWER INTERNATIONAL SERIES
IN ENGINEERING AND COMPUTER SCIENCE

KNOWLEDGE DISCOVERY
AND MEASURES OF INTEREST

by

Robert J. Hilderman
University of Regina, Canada

Howard J. Hamilton
University of Regina, Canada

KLUWER ACADEMIC PUBLISHERS
Boston / Dordrecht / London

Distributors for North, Central and South America:
Kluwer Academic Publishers
101 Philip Drive
Assinippi Park
Norwell, Massachusetts 02061 USA
Telephone (781) 871-6600
Fax (781) 871-6528
E-Mail <kluwer@wkap.com>

Distributors for all other countries:
Kluwer Academic Publishers Group
Distribution Centre
Post Office Box 322
3300 AH Dordrecht, THE NETHERLANDS
Telephone 31 78 6392 392
Fax 31 78 6546 474
E-Mail <services@wkap.nl>

 Electronic Services <http://www.wkap.nl>

Library of Congress Cataloging-in-Publication Data

Hilderman, Robert J.
  Knowledge discovery and measures of interest/by Robert J. Hilderman, Howard J. Hamilton.
    p. cm. – (The Kluwer international series in engineering and computer science;SECS 638)
  Includes bibliographical references and index.

  1. Data mining. 2. Database searching. 3. Expert systems (Computer science). I.
Hamilton, Howard J. II. Title. III. Series.

QA76.9.D343 H56 2001
006.3—dc21

ISBN 978-1-4419-4913-4   2001038585

Contents

List of Figures		ix
List of Tables		xi
Preface		xv
Acknowledgments		xix
1. INTRODUCTION		1
1.1	KDD in a Nutshell	1
	1.1.1 The Mining Step	2
	1.1.2 The Interpretation and Evaluation Step	7
1.2	Objective of the Book	9
2. BACKGROUND AND RELATED WORK		11
2.1	Data Mining Techniques	11
	2.1.1 Classification	11
	2.1.2 Association	12
	2.1.3 Clustering	13
	2.1.4 Correlation	14
	2.1.5 Other Techniques	15
2.2	Interestingness Measures	15
	2.2.1 Rule Interest Function	15
	2.2.2 *J*-Measure	16
	2.2.3 Itemset Measures	16
	2.2.4 Rule Templates	17
	2.2.5 Projected Savings	17
	2.2.6 *I*-Measures	18
	2.2.7 Silbershatz and Tuzhilin's Interestingness	18
	2.2.8 Kamber and Shinghal's Interestingness	19
	2.2.9 Credibility	20
	2.2.10 General Impressions	20
	2.2.11 Distance Metric	21

2.2.12	Surprisingness	21
2.2.13	Gray and Orlowska's Interestingness	22
2.2.14	Dong and Li's Interestingness	22
2.2.15	Reliable Exceptions	23
2.2.16	Peculiarity	23
3.	**A DATA MINING TECHNIQUE**	**25**
3.1	Definitions	25
3.2	The Serial Algorithm	26
	3.2.1 General Overview	26
	3.2.2 Detailed Walkthrough	28
3.3	The Parallel Algorithm	30
	3.3.1 General Overview	31
	3.3.2 Detailed Walkthrough	32
3.4	Complexity Analysis	33
	3.4.1 Attribute-Oriented Generalization	33
	3.4.2 The All_Gen Algorithm	33
3.5	A Comparison with Commercial OLAP Systems	34
4.	**HEURISTIC MEASURES OF INTERESTINGNESS**	**37**
4.1	Diversity	37
4.2	Notation	39
4.3	The Sixteen Diversity Measures	39
	4.3.1 The $I_{Variance}$ Measure	39
	4.3.2 The $I_{Simpson}$ Measure	40
	4.3.3 The $I_{Shannon}$ Measure	40
	4.3.4 The I_{Total} Measure	41
	4.3.5 The I_{Max} Measure	41
	4.3.6 The $I_{McIntosh}$ Measure	42
	4.3.7 The I_{Lorenz} Measure	42
	4.3.8 The I_{Gini} Measure	43
	4.3.9 The I_{Berger} Measure	44
	4.3.10 The I_{Schutz} Measure	44
	4.3.11 The I_{Bray} Measure	44
	4.3.12 The $I_{Whittaker}$ Measure	44
	4.3.13 The $I_{Kullback}$ Measure	45
	4.3.14 The $I_{MacArthur}$ Measure	45
	4.3.15 The I_{Theil} Measure	46
	4.3.16 The $I_{Atkinson}$ Measure	46

5. AN INTERESTINGNESS FRAMEWORK   47
    5.1   Interestingness Principles   47
    5.2   Summary   49
    5.3   Theorems and Proofs   51
       5.3.1   Minimum Value Principle   51
       5.3.2   Maximum Value Principle   63
       5.3.3   Skewness Principle   79
       5.3.4   Permutation Invariance Principle   84
       5.3.5   Transfer Principle   84

6. EXPERIMENTAL ANALYSES   99
    6.1   Evaluation of the All_Gen Algorithm   99
       6.1.1   Serial vs Parallel Performance   100
       6.1.2   Speedup and Efficiency Improvements   103
    6.2   Evaluation of the Sixteen Diversity Measures   104
       6.2.1   Comparison of Assigned Ranks   105
       6.2.2   Analysis of Ranking Similarities   107
       6.2.3   Analysis of Summary Complexity   112
       6.2.4   Distribution of Index Values   117

7. CONCLUSION   123
    7.1   Summary   123
    7.2   Areas for Future Research   125

Appendices   141
    Comparison of Assigned Ranks   141
    Ranking Similarities   149
    Summary Complexity   155

Index   161

List of Figures

1.1	A DGG for the *Office* attribute	3
1.2	A multi-path DGG for the *Office* attribute	5
1.3	DGGs for the *Shape*, *Size*, and *Colour* attributes	7
1.4	Which summary should be considered most interesting?	8
3.1	Serial multi-attribute generalization algorithm	27
3.2	Parallel multi-attribute generalization algorithm	31
4.1	A sample Lorenz curve	43
6.1	Relative performance generalizing two attributes	101
6.2	Relative performance generalizing three attributes	102
6.3	Relative performance generalizing four attributes	102
6.4	Relative performance generalizing five attributes	103
6.5	Relative complexity of summaries within *N-3*	115
6.6	Relative complexity of summaries within *C-4*	115
6.7	Relative complexity of summaries between NSERC discovery tasks	116
6.8	Relative complexity of summaries between Customer discovery tasks	116
6.9	Histogram of index value frequencies for $I_{Variance}$	118
6.10	Histogram of index value frequencies for I_{Schutz}	118

List of Tables

1.1	A sales transaction database	4
1.2	An example sales summary	4
1.3	Domains for the *Shape*, *Size*, and *Colour* attributes	6
1.4	Domain for the compound attribute *Shape-Size-Colour*	6
1.5	Summary for the DGG node combination *ANY-Package-Colour*	7
3.1	Summary for the DGG node combination *Shape-Size-ANY*	29
3.2	Summary for the DGG node combination *Shape-Package-Colour*	29
3.3	Summary for the DGG node combination *Shape-Package-ANY*	30
3.4	A sample dimension map for the *Shape*, *Size*, and *Colour* attributes	35
4.1	A sample summary	39
5.1	Measures satisfying the principles (concentration order and dispersion order)	49
5.2	Measures satisfying the principles (aggregate order)	50
6.1	Characteristics of the DGGs associated with the selected attributes	100
6.2	Speedup and efficiency results obtained using the parallel algorithm	104
6.3	Ranks assigned by $I_{Variance}$ and $I_{Simpson}$ from *N-2*	106
6.4	Summary 1 from *N-2*	107
6.5	Ranking similarities for NSERC discovery tasks	109
6.5	Ranking similarities for NSERC discovery tasks (continued)	110
6.5	Ranking similarities for NSERC discovery tasks (continued)	111
6.6	Relative interestingness versus complexity for NSERC discovery tasks	112

6.6	Relative interestingness versus complexity for NSERC discovery tasks (continued)	113
6.7	Ordered arrangements of two populations	117
6.8	Skewness and kurtosis of the index values for the two populations	119
6.9	Distribution of index values for 50 objects among 10 classes	120
6.10	Distribution of index values for 50 objects among 5 classes	120
6.11	Vectors at the middle index value for two populations	121
A.1	Ranks assigned by $I_{Shannon}$ and I_{Total} from N-2	142
A.2	Ranks assigned by I_{Max} and $I_{McIntosh}$ from N-2	143
A.3	Ranks assigned by I_{Lorenz} and I_{Berger} from N-2	144
A.4	Ranks assigned by I_{Schutz} and I_{Bray} from N-2	145
A.5	Ranks assigned by $I_{Whittaker}$ and $I_{Kullback}$ from N-2	146
A.6	Ranks assigned by $I_{MacArthur}$ and I_{Theil} from N-2	147
A.7	Ranks assigned by $I_{Atkinson}$ and I_{Gini} from N-2	148
B.1	Ranking similarities for Customer discovery tasks	151
B.1	Ranking similarities for Customer discovery tasks (continued)	152
B.1	Ranking similarities for Customer discovery tasks (continued)	153
C.1	Relative interestingness versus complexity for *C-2* and *C-3*	157
C.1	Relative interestingness versus complexity for *C-2* and *C-3* (continued)	158
C.2	Relative interestingness versus complexity for *C-4* and *C-5*	159
C.2	Relative interestingness versus complexity for *C-4* and *C-5* (continued)	160

Preface

During the last two decades, the capability for collecting and storing data has grown as database and storage technology has become more advanced and cost effective. Consequently, many organizations began, and continue, to archive vast amounts of data because it is assumed that useful knowledge can be extracted from the data once it is analyzed. However, early in the last decade it was realized that our ability to collect and store data was beginning to far exceed our ability to efficiently analyze it. To address this problem, researchers from statistics, artificial intelligence, pattern recognition, machine learning, databases, and data visualization began to develop tools for the intelligent and automatic discovery of knowledge in databases. The resulting body of work and research came to be known as knowledge discovery in databases.

Knowledge discovery in databases, also commonly known as data mining, is universally considered to be the non-trivial process of identifying previously unknown, valid, novel, potentially useful, and understandable patterns in data. It encompasses many different techniques that differ in the kind of data that can be analyzed and the form of knowledge representation used to convey the discovered patterns. Typically, the number of patterns generated is very large, but only a few of these patterns are likely to be of any interest to the domain expert analyzing the data. The reason for this is that many of the patterns are either irrelevant, or obvious, and do not provide any new knowledge. To increase the utility, relevance, and usefulness of the discovered patterns, techniques are required to reduce the number of patterns that need to be considered and to rank those that are likely to be most interesting. Techniques that satisfy this goal are broadly referred to as interestingness measures.

In this book, we study two closely related steps in any knowledge discovery system: the generation of discovered knowledge, and the interpretation and evaluation of the discovered knowledge. In the generation step, we study data summarization, where a single dataset can be generalized in many different ways and to many levels of granularity according to a hierarchical data structure called a domain generalization graph. A domain generalization graph is associated with an attribute in a database and is a directed graph, where each node represents a different way of summarizing the possible domain values associated with the attribute, and each edge represents a generalization relation between adjacent domains. In the interpretation and evaluation step, we study diversity measures as heuristic measures of interestingness for ranking the summaries created in the generation step. The tuples in a summary are unique, and therefore, can be considered to be a population with a structure that can be described by some frequency or probability distribution. The diversity measures used in this work operate on these frequency or probability distributions to generate a single numeric value that can be used to rank the interestingness of each summary relative to the other summaries generated from the database in the same discovery task. Although, diversity measures have seen extensive use in the physical, social, ecological, management, information, and computer sciences, their use for ranking summaries generated from databases is a natural and useful extension into a new application domain.

The book is designed to provide both knowledge discovery researchers and practitioners with the background necessary for the selection and application of interestingness measures in knowledge discovery systems. The knowledge discovery researcher will find that the material provides a theoretical foundation for interestingness in data mining applications where diversity measures are used to rank summaries. The theoretical foundation provides the basis for an intuitive understanding of the term "interestingness" when used within this context. Similarly, the knowledge discovery practitioner will find solid empirical evidence on which to base decisions regarding the choice of potential measures. That is, when choosing any candidate interestingness measure for ranking summaries, the practitioner will be better able to judge the suitability of the candidate interestingness measure for the intended application. Thus, given the strong theoretical and empirical nature of the material, both researchers and practitioners can benefit from reading the book.

The reader should have some knowledge of the basic concepts and terminology associated with database systems. In addition, some background in elementary statistics and machine learning may also be useful, but is not necessarily required, as the concepts and techniques discussed within the book can be utilized without knowledge of the underlying theory or processes.

The book consists of seven chapters. Chapter 1 provides a brief introduction to the general framework of knowledge discovery in databases, and positions our work within this framework via a broad overview of the algorithms, concepts, and techniques utilized to generate and rank discovered knowledge.

Chapter 2 presents a general overview of classical data mining techniques and algorithms, highlighting the significant characteristics of each technique.

A detailed survey of relevant interestingness measures is also presented to highlight important developments in the area of interestingness measures.

Chapter 3 introduces the conceptual model for domain generalization graphs and defines our notion of summaries. Serial and parallel versions of our algorithm for efficiently generating summaries according to the domain generalization graphs associated with a set of attributes is also presented.

Chapter 4 describes various measures of diversity that we propose as heuristic measures for ranking the interestingness of summaries generated from databases.

Chapter 5 develops a theory of interestingness through the mathematical formulation of five principles that must be satisfied by any acceptable measure of interestingness used for ranking summaries generated from databases. Theoretical results describe, through mathematical proof, those measures that satisfy the proposed principles.

Chapter 6 summarizes the performance of the serial and parallel summary generation algorithms, the results obtained from a variety of discovery tasks run against industrial databases. It also characterizes the behaviour of the proposed diversity measures when used to rank the interestingness of summaries generated from synthetic data.

Chapter 7 provides a summary of our work and suggests areas for future research.

ROBERT J. HILDERMAN
HOWARD J. HAMILTON

Acknowledgments

We acknowledge the support of the Institute for Robotics and Intelligent Systems, the Networks of Centres of Excellence Program of the Government of Canada, the Natural Sciences and Engineering Research Council of Canada (NSERC), and the participation of PRECARN Associates, Inc., Canadian Cable Labs, Inc., and the University of Regina.

We thank Dr. Guy Mineau, Dr. Yiyu Yao, Dr. Nick Cercone, and Dr. Gemai Chen for their comments, suggestions, and criticisms.

We also thank Kluwer Academic Publishers, particularly Lance Wobus and Sharon Palleschi, for making this book possible.

Chapter 1

INTRODUCTION

This book is about generating patterns from databases and then applying measures to the patterns to determine their relative interestingness. In particular, the book describes a technique for mining summaries from databases using domain generalization graphs and objective measures of interestingness. The technique utilized here is included in the broad set of techniques encompassed by the discipline known as knowledge discovery in databases (KDD).

1.1. KDD in a Nutshell

Knowledge discovery in databases, also known as *data mining*, has been universally accepted to be the non-trivial process of identifying previously unknown, valid, novel, potentially useful, and understandable patterns in databases [33, 37]. The term *non-trivial* implies that the process is not a simple one and that it likely consumes some significant amount of computer resources in terms of storage and/or processor capacity, possibly over an extended period of time. The term *previously unknown* implies that the discovered patterns represent new knowledge. The term *valid* implies that the discovered patterns are reproducible and would likely be consistent with patterns discovered in similar data from other sources. The term *novel* implies that the knowledge being sought is not obvious or intuitive. The term *potentially useful* implies that the knowledge provide some insight, have some applicability to solving real-world problems, and aid in decision-making processes. The term *understandable* implies that the results be presented in a manner that is suitable for human consumption.

The process of KDD typically includes many steps, but at a minimum, the following steps are usually required (not necessarily in the order listed) [33, 60].

- The *cleaning and preprocessing* step attempts to eliminate errors, omissions, and inconsistencies through data hygiene operations such as reconstructing damaged or missing data, and standardizing the use of abbreviations.

- The *integration* step combines data from multiple sources.

- The *selection* step relies upon tools such as an SQL or natural language interface or query-by-example template to describe the data that we want to retrieve from the database.

- The *reduction and transformation* step attempts to reduce the volume of data that needs to be considered in subsequent steps by converting the data into a format more amenable to mining by the selected algorithm.

- The *mining* step specifies a task to be performed, such as applying some summarization or anomaly-detection algorithm to the data.

- The *interpretation and evaluation* step attempts to understand the results of the data mining step by identifying deviations from the expected results or previously established norms, or by quantifying the degree to which a result is surprising or interesting.

- The *presentation* step utilizes visualization and knowledge representation techniques to present the results in a manner suitable for human understanding.

- The *application* step uses the newly discovered and understood knowledge as part of some problem-solving or decision-making process.

- The *refinement and repetition* step attempts to use the newly discovered knowledge to improve the quality or focus of the discovery task to make it more relevant to the problem-solving or decision-making process.

In this book, we focus on the mining step and the interpretation and evaluation step.

1.1.1.   The Mining Step

Data mining algorithms can be broadly classified into two general areas: summarization and anomaly detection [71]. *Summarization algorithms* find concise descriptions of input data. For example, *classificatory algorithms* partition input data into disjoint groups. The results of such classification might be represented as a high-level summary, a decision tree, or a set of characteristic rules, as with C4.5 [112], DBLearn [58], and KID3 [110]. *Anomaly-detection algorithms* identify unusual features of data, such as combinations that occur with greater or lesser frequency than might be expected. For example, *association algorithms* find, from transaction records, sets of items that appear with each other in sufficient frequency to merit attention [2, 17, 18, 106, 134]. Similarly, *sequencing algorithms* find relationships among items or events across time, such as events A and B usually precede event C [3, 7, 130]. Hybrid approaches that generate high-level association rules from input data and concept hierarchies have also been investigated [57, 63, 129].

Here we introduce a new data mining technique based upon a data structure called a *domain generalization graph* (DGG), and an algorithm that generates from a database all possible summaries described by a DGG. A DGG defines a partial order representing a set of generalization relations for an attribute. A sample DGG for the *Office* attribute in a sales database is shown in Figure 1.1. In Figure 1.1, each node is a partition of the domain values that can be used to describe the attribute and the edges connecting each pair of nodes defines a generalization relation based upon a summarization technique known as *attribute-oriented generalization* (AOG) [54, 55, 56].

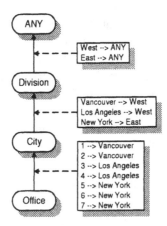

Figure 1.1. A DGG for the *Office* attribute

AOG summarizes the information in a database by replacing specific attribute values with more general concepts according to user-defined taxonomies. For example, the domain for the *Office* attribute is represented by the *Office* node. Increasingly more general descriptions of the domain values are represented by the *City*, *Division*, and *ANY* nodes. The edges between adjacent nodes are each associated with a generalization relation. In Figure 1.1, the generalization relation consists of a table lookup (other generalization relations besides table lookups are possible [51, 115], but we restrict our discussion to table lookups for the sake of simplicity and clarity). The table associated with the edge between the *Office* and *City* nodes defines the mapping of the domain values of the *Office* node to the domain values of the *City* node. That is, the values 1 and 2 are mapped to *Vancouver*, 3 and 4 are mapped to *Los Angeles*, and 5, 6, and 7 are mapped to *New York*. The table associated with the edge between the

City and *Division* nodes can be described similarly. The table associated with the edge between the *Division* and *ANY* nodes maps all values in the *Division* domain to the special value *ANY*. So, the domain values for the *Office* node correspond to the most specific representation of the data, the domain values for the *City* and *Division* nodes correspond to a more general representation of the data, and the *ANY* node corresponds to the most general representation of the data.

Given the sales transaction database shown in Table 1.1, one of the many possible summaries that can be generated is shown in Table 1.2. For each tuple in the sales transaction database, we simply substitute the appropriate value from the lookup table associated with the edge being traversed, and keep a count of the number of tuples that have been aggregated from the original unconditioned data in a derived attribute called *Count*. We also aggregate any numeric attributes that have been selected for inclusion in the summaries being generated (here we assume the *Office*, *Quantity*, and *Amount* attributes have been selected). These steps are repeated as each edge in the DGG is traversed, resulting in the generation of a new and unique summary following each traversal. For example, in Figure 1.1, the edges between the *Office* and *City* nodes and the *City* and *Division* nodes have been traversed, in sequence, resulting in the generation of the summary shown in Table 1.2, and corresponding to the *Division* node. The generalization space (i.e., the set of all possible summaries that can be generated) for the DGG shown in Figure 1.1 consists of $4 - 1 = 3$ summaries, those corresponding to the *City*, *Division*, and *ANY* nodes.

Table 1.1.   A sales transaction database

Office	Shape	Size	Colour	Quantity	Amount
2	round	small	white	2	$50.00
5	square	medium	black	3	$75.00
3	round	large	white	1	$25.00
7	round	large	black	4	$100.00
1	square	x-large	white	3	$75.00
6	round	small	white	4	$100.00
4	square	small	black	2	$50.00

Table 1.2.   An example sales summary

Office	Quantity	Amount	Count
West	8	$200.00	4
East	11	$275.00	3

When there are multiple DGGs associated with an attribute, meaning knowledge about the attribute can be expressed in different ways, a multi-path DGG
can be constructed from the single-path DGGs. For example, the DGG shown
on the right side of Figure 1.2 is a multi-path DGG that has been constructed
from the two single-path DGGs shown on the left. There are now two possible
edges that can be traversed from the *City* node, from which summaries can be
generated corresponding to the *Division* and *Country* nodes. Here we assume
that common names used to describe the nodes in the single-path DGGs associated with an attribute represents the same partition of the domain, and that the
edge connecting adjacent nodes describes the same generalization relation. For
example, in Figure 1.2, since we assume that the *ANY*, *City*, and *Office* nodes in
the single-path DGGs on the left represent the same partition of the domain and
that the edges describe the same generalization relation, then the like-named
nodes and edges can be combined in the multi-path DGG. The generalization
space for the multi-path DGG consists of $5 - 1 = 4$ summaries, corresponding
to the *City*, *Division*, *Country*, and *ANY* nodes. In contrast, the generalization
space for the two single path DGGs consists of $2 \times (4 - 1) = 6$ summaries,
containing duplicates of those summaries corresponding to the *City* and *ANY*
nodes.

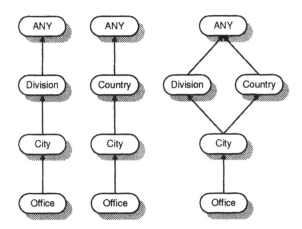

Figure 1.2. A multi-path DGG for the *Office* attribute

So far, we have only been concerned with the summaries generated from a
database where a single attribute is associated with a DGG. Of course, we may
want to generate summaries from databases where multiple attributes are associated with DGGs. In this situation, known as multi-attribute generalization,
a set of individual attributes can be considered to be a single attribute (called
a compound attribute) whose domain is the cross-product of the individual
attribute domains. For example, given the domains for the individual attributes

Shape, *Size*, and *Colour* shown in Table 1.3, the domain for the compound attribute *Shape-Size-Colour* is as shown in Table 1.4.

Table 1.3. Domains for the *Shape*, *Size*, and *Colour* attributes

Shape	Size	Colour
round	small	black
square	medium	white
	large	
	x-large	

Table 1.4. Domain for the compound attribute *Shape-Size-Colour*

Shape-Size-Colour
round-small-black
round-small-white
round-medium-black
round-medium-white
round-large-black
round-large-white
round-x-large-black
round-x-large-white
square-small-black
square-small-white
square-medium-black
square-medium-white
square-large-black
square-large-white
square-x-large-black
square-x-large-white

A summary generated from the cross-product domain for the compound attribute *Shape-Size-Colour* corresponds to a unique combination of nodes from the DGGs associated with the individual attributes, where one node is selected from the DGG associated with each attribute. For example, given the sales transaction database shown in Table 1.1 (assume the *Shape*, *Size*, and *Colour* attributes have been selected for generalization) and the associated DGGs shown in Figure 1.3, one of the many possible summaries that can be generated is shown in Table 1.5. The summary in Table 1.5 is obtained by generalizing the *Shape* attribute to the *ANY* node and the *Size* attribute to the *Package* node, while the *Colour* attribute remains ungeneralized.

The complexity of the DGGs is a primary factor determining the number of summaries that can be generated, and depends only upon the number of

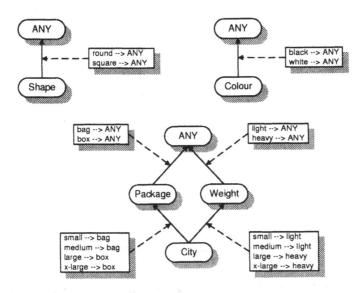

Figure 1.3. DGGs for the *Shape*, *Size*, and *Colour* attributes

Table 1.5. Summary for the DGG node combination *ANY-Package-Colour*

Shape	Size	Colour	Count
ANY	bag	white	2
ANY	bag	black	2
ANY	box	white	2
ANY	box	black	1

nodes in the generalization space, it is not dependent upon the number of tuples in the original input relation. For example, the generalization space for discovery tasks involving the *Shape*, *Size*, and *Colour* attributes consists of $(2 \times 4 \times 2) - 1 = 15$ summaries.

1.1.2. The Interpretation and Evaluation Step

When the number of attributes to be generalized is large or the DGGs associated with a set of attributes are complex, many summaries may be generated. A user may then be required to evaluate each one to determine whether it contains an interesting result. For example, consider the two summaries shown in Figure 1.4. Which summary should be considered to be the most interesting? Perhaps given just two summaries, this is not such a daunting task. But given hundreds, or even thousands of summaries, it is simply not feasible to consider trying to identify all of the interesting summaries using a manual technique.

Colour	Shape	Count
red	round	40
green	round	20
red	square	20
blue	square	20

Colour	Shape	Count
red	box	33
red	bag	33
green	bag	20
blue	box	14

Figure 1.4.   Which summary should be considered most interesting?

What is needed is an effective measure of interestingness to assist in the interpretation and evaluation of the discovered knowledge. The development of such measures is currently an active research area in KDD. Such measures are broadly classified as either objective or subjective. *Objective measures* are based upon the structure of discovered patterns, such as the frequency with which combinations of items appear in sales transactions [2, 6]. *Subjective measures* are based upon user beliefs or biases regarding relationships in the data, such as an approach utilizing Bayes Rule to revise prior beliefs [105], or an approach utilizing templates to describe interesting patterns [82]. In [68], we describe principles that guide the use of objective interestingness measures, and techniques are described in [73] for visualizing discovered knowledge using objective interestingness measures.

In this book, we study sixteen diversity measures and evaluate their applicability as heuristics for ranking the interestingness of summaries generated from databases. The measures are all well-known measures of dispersion, dominance, inequality and concentration that have previously been successfully applied in several areas of the physical, social, ecological, management, information, and computer sciences. We restrict our study to objective measures of interestingness that consider only the frequency or probability distribution of the values in the derived *Count* attribute of the summaries generated. For example, statistical variance, σ^2, is one of the diversity measures evaluated. Based upon the probability distribution of the values contained in the derived *Count* attribute for the summaries in Figure 1.4 (i.e., each value in the *Count* attribute is converted to a percentage of the total), $\sigma^2 = 0.01$ and $\sigma^2 = 0.0091$ for the first and second summaries, respectively. If we consider the most interesting summary to simply be that which has the highest variance, then the first summary is most interesting.

Although our measures were developed and utilized for ranking the interestingness of summaries (i.e., generalized relations) using DGGs, they are more generally applicable to other problem domains. For example, alternative methods could be used to guide the generation of summaries, such as Galois lattices [41], conceptual graphs [15], or formal concept analysis [132]. Also, summaries could more generally include structures such as *views* (i.e., precomputed, virtual tables derived from a relational database), *data cubes* (i.e., several multidimensional databases representing aggregated data from a data warehouse) [45], or *summary tables* (i.e., materialized, aggregate views derived from a data cube). The application of our technique to ranking knowledge contained in data cubes and characterized association rules is described in [67] and [69].

1.2.   Objective of the Book

The objective of this book is to introduce and evaluate a technique for ranking the interestingness of discovered patterns in data. Realizing this objective consists of four primary goals:

- To introduce DGGs for describing and guiding the generation of summaries from databases.

- To introduce and evaluate serial and parallel algorithms that traverse the generalization space described by the DGGs.

- To introduce and evaluate diversity measures as heuristic measures of interestingness for ranking summaries generated from databases.

- To develop the preliminary foundation for a theory of interestingness within the context of ranking summaries generated from databases.

Chapter 2

BACKGROUND AND RELATED WORK

In KDD, the knowledge that we seek to discover describes patterns in the data as opposed to knowledge about the data itself. Patterns in the data can be represented in many different forms, including classification rules, association rules, clusters, sequential patterns, time series, contingency tables, summaries obtained using some hierarchical or taxonomic structure, and others. Typically, the number of patterns generated is very large, but only a few of these patterns are likely to be of any interest to the domain expert analyzing the data. The reason for this is that many of the patterns are either irrelevant or obvious, and do not provide new knowledge [105]. To increase the utility, relevance, and usefulness of the discovered patterns, techniques are required to reduce the number of patterns that need to be considered. Techniques which satisfy this goal are broadly referred to as interestingness measures.

2.1. Data Mining Techniques

Data mining encompasses many different techniques and algorithms. These differ in the kinds of data that can be analyzed and the kinds of knowledge representation used to convey the discovered knowledge. Here we describe some of the more successful and widely known techniques and algorithms.

2.1.1. Classification

Classification is perhaps the most commonly applied data mining technique. Early examples of classification techniques from the literature include VSA [100, 101], ID3 [113], AQ15 [98], and CN2 [27]. VSA induces a single classification rule from two complementary trees (a specialization tree and a generalization tree) that converge on a common node containing the rule. ID3 induces a decision tree. An object is classified by descending the tree until a branch leads to a leaf node containing the decision. AQ15 induces a set of decision rules. An object is classified by selecting the most preferred decision rule according to user-defined criteria. CN2 induces a decision list. An object is classified by selecting the best rule according to user-defined accuracy and statistical significance criteria.

Later examples of classification techniques from the literature include FCLS [143], PrIL [47], SLIQ [97], and CLOUDS [11]. FCLS induces a weighted threshold rule. The threshold determines the number of conditions which must be satisfied in a valid rule. An object is classified by generalizing and specializing examples until the number of incorrectly classified examples is below some user-defined error rate. PrIL induces decision rules in a manner similar to those induced by ID3. However, the rules induced by PrIL are associated with a minimum correct classification threshold and confidence level. When a rule cannot meet the minimum correct classification threshold, objects cannot be classified according to that rule. SLIQ induces a decision tree built using the Minimum Description Length principle [119]. It is similar to other decision tree classifiers except that it is capable of handling large disk-resident datasets (i.e., all of the data cannot fit into memory). CLOUDS induces a decision tree similar to the manner used by SLIQ except that a more computationally efficient method is used to determine the splitting points at each node.

Other examples of classification techniques from the literature include C4.5/C5.0 [112], KID3 [110], parallel ID3 [35], and SPRINT [124]. C4.5/C5.0 is an industrial-quality descendant of ID3 that has seen widespread use in the research community. KID3 induces exact decision rules (i.e., those that are always correct) and strong decision rules (i.e., those that are almost always correct). An efficient parallel technique is used that accesses the data only once to generate all exact rules. Parallel ID3 uses a distributed tree construction technique to induce decision trees. SPRINT is a parallel version of the SLIQ algorithm that uses different and more memory efficient data structures to induce a decision tree.

2.1.2.   Association

Association is another of the commonly applied data mining techniques. The problem is typically examined in the context of discovering buying patterns from retail sales transactions, and is commonly referred to as market basket analysis. Market basket analysis was originally introduced in [2] and has since been studied extensively.

Much of the literature focuses on the Apriori algorithm [4] and its descendants containing various refinements. Apriori extracts the set of frequent itemsets from the set of candidate itemsets generated. A frequent itemset is an itemset whose support is greater than some user-defined minimum and a candidate itemset is an itemset whose support has yet to be determined. It has an important property that if any subset of a candidate itemset is not a frequent itemset, then the candidate itemset is also not a frequent itemset.

Refinements to Apriori include Partition [122], DHP [106], sampling [134], DIC [18], and parallel Apriori [5]. Partition reads the database at most two times to generate all significant association rules, while generating no false negatives. It is also inherently parallel in nature and can be parallelized with minimal communication and synchronization between nodes. DHP is a hash-based algorithm for generating candidate itemsets that reduces the number of candidate 2-itemsets by an order of magnitude. Pruning the candidate 2-itemsets significantly reduces the number of frequent k-itemsets that need to be considered when $k > 2$. Sampling is used to take a random sample from a database to find all association rules that are probably valid in the entire database. A second pass of the database is used to verify the support for each potential association rule. DIC partitions the database into blocks. When scanning the first block it counts only 1-itemsets. When scanning the k-th block, it counts 2-, 3-, 4-, ..., k-itemsets. Usually it can finish counting all the itemsets in two passes over the data. Parallel Apriori is a parallel version of Apriori that exhibits excellent scaleup behaviour and requires only minimal additional overhead compared to serial Apriori.

Other literature focuses on alternative approaches for discovery of association rules. These approaches include Hybrid Distribution [53], Itemset Clustering [141], share measures [64], and Q2 [19]. Hybrid Distribution is a parallel algorithm that improves upon parallel Apriori by dynamically partitioning the candidate itemsets to achieve superior load balancing across the nodes. More association rules can then be generated more quickly in a single pass over the database. Scaleup is near linear and in general it utilizes memory more efficiently. Itemset Clustering approximates the set of potentially maximal frequent itemsets and then uses an efficient lattice traversal technique to generate clusters of frequent itemsets in a single pass over the database. Share measures are used to more accurately indicate the financial impact of an itemset by not only considering the co-occurrence of items in an itemset, but by also considering the quantity and value of the items purchased. Q2 obtains performance improvements of more than an order of magnitude over Apriori by computing and pruning the frequent Boolean itemsets before searching for valid association rules. Once this is done, association rules can be found with a single pass over the database.

2.1.3. Clustering

Identifying objects that share some distinguishing characteristics is also a frequently used data mining technique. Known as clustering, there are numerous techniques described in the literature.

Early examples of clustering from the literature include CLUSTER/2 [99] and COBWEB [36]. CLUSTER/2 finds a disjoint clustering of objects that optimizes user-defined parameters regarding the number of clusters required and clustering quality criteria. It uses an efficient path-rank-ordered search procedure to limit the number of nodes visited in the search tree. COBWEB further increases efficiency by using an incremental approach that organizes data in a way that maximizes its inference abilities by identifying data dependencies involving important attributes.

More recent examples from the literature include CLARANS [103], BIRCH [144], DBSCAN [32], STING [135], and CLIQUE [1]. CLARANS is an extension of the k-medoids approaches developed in [80]. It is based upon a randomized search algorithm with user-defined parameters that control the length and quality of the search. BIRCH incrementally and dynamically evaluates data to generate the best quality clusters possible given user-defined time constraints and available memory. A single pass over the database is usually enough to find high quality clusters. DBSCAN is a density-based approach that utilizes user-defined parameters for controlling the density of the discovered clusters. This approach allows adjacent regions of sufficiently high density to be connected to form clusters of arbitrary shape and is able to differentiate noise in regions of low density. STING models the search space as a hierarchical structure of rectangular cells corresponding to different levels of resolution. Each cell at a high level is partitioned to form a number of smaller cells in the next lower level. Statistical information is associated with each cell to facilitate querying and incremental updates. CLIQUE is a density-based approach that has the ability to find clusters in subspaces of high dimensional data. The search space is partitioned into equal-sized units. Discovered clusters are unions of adjacent high-density units.

2.1.4.   Correlation

Statistically-oriented in nature, correlation has seen increasing use as a data mining technique. Although the analysis of multi-dimensional categorical data is possible and described extensively in the literature [34, 42, 118], the most commonly employed method is that of two-dimensional contingency table analysis of categorical data using the chi-square statistic as a measure of significance.

There are many recent examples in the literature [17, 83, 89, 121, 142]. In [121], contingency tables are analyzed to discover students who are poorly prepared for university level course work and at risk of dropping out. In [83], contingency tables are analyzed to discover simple associations between single attributes that can be easily visualized in a bar graph. In [142], contingency tables are analyzed using the 49er data mining system. 49er examines each pair of attributes in a contingency table and applies statistical test of significance and strength to quantify the discovered patterns. In [17], contingency tables are analyzed to generate dependence rules that identify statistical dependence in both the presence and absence of items in itemsets. And in [89], contingency tables are analyzed to discover unexpected and interesting patterns that have a low level of support and a high level of confidence.

2.1.5. Other Techniques

Other data mining techniques search for patterns in sequences and time series. The problem of mining for patterns in sequences was introduced in [7, 130]. The search for sequences of events that occur in a particular order and within a particular time interval is described in [93, 94]. A logic for expressing temporal patterns defined over categorical data as a means for discovering patterns in sequences is described in [104]. Recent approaches for the discovery of patterns in sequences are described in [48, 136, 140].

The problem of mining for patterns in time series has received a considerable amount of attention recently. An approach that queries the Fourier series representation of a sequence is described in [114]. A framework is described in [116] where models containing high-dimensional time series data are learned so that data values can be forecast for the immediate future. An extended representation of time series that allows accurate classification and clustering through a relevance feedback mechanism is described in [81]. A method for mining segment-wise periodicity in time series data is described in [59]. The problem of finding rules relating patterns in a time series to other patterns in the same time series, or to patterns in another time series are described in [29].

2.2. Interestingness Measures

One problem area in KDD is the development of interestingness measures for ranking the usefulness and utility of discovered patterns. In this section, we survey and describe significant interestingness measures described in the literature.

2.2.1. Rule-Interest Function

The rule-interest function [110] is used to quantify the correlation between attributes in a simple classification rule. A simple classification rule is one where the left- and right-hand sides of the logical implication $X \rightarrow Y$ correspond to a single attribute. The *rule-interest function* is given by

$$RI = |X \cap Y| - \frac{|X||Y|}{N},$$

where N is the total number of tuples, $|X|$ and $|Y|$ are the number of tuples satisfying conditions X and Y, respectively, $|X \cap Y|$ is the number of tuples satisfying $X \rightarrow Y$, and $|X||Y|/N$ is the number of tuples expected if X and Y were independent (i.e., not associated).

When $RI = 0$, then X and Y are statistically independent and the rule is not interesting. When $RI > 0$ ($RI < 0$), then X is positively (negatively) correlated to Y. The significance of the correlation between X and Y can be determined using the chi-square test for a 2×2 contingency table. Those rules which do not exceed a predetermined minimum significance threshold are determined to be the most interesting.

2.2.2.   J-Measure

The J-measure [128] is the average information content of a probabilistic classification rule and is used to find the best rules relating discrete-valued attributes. A probabilistic classification rule is a logical implication $X \rightarrow Y$ with some probability p, where the left- and right-hand sides correspond to a single attribute. The right-hand side is restricted to simple single-valued assignment expressions, while the left-hand side may be a conjunction of these simple expressions. The *J-measure* is given by

$$J(x; y) = p(y) \left[p(x|y) \log \left(\frac{p(x|y)}{p(x)} \right) + (1 - p(x|y)) \log \left(\frac{1 - p(x|y))}{(1 - p(x))} \right) \right],$$

where $p(y)$, $p(x)$, and $p(x|y)$ are the probabilities of occurrence of y, x, and x given y, respectively, and the term inside the square brackets is the cross entropy. Cross entropy is the measure of the goodness of fit of two probability distributions.

High values for $J(x; y)$ are desirable, but are not necessarily associated with the best rule. For example, rare conditions may be associated with the highest values for $J(x; y)$ (i.e., where a particular y is highly unlikely), but the resulting rule is insufficiently general to provide any new information. Consequently, analysis may be required in which the accuracy of a rule is traded for some level of generality or goodness-of-fit.

2.2.3.   Itemset Measures

The itemset measures [2, 6], are used to identify frequently occurring association rules from sets of items in large databases. An association rule is a logical implication $X \rightarrow Y$, where the left- and right-hand sides correspond to a set of attributes, and X and Y are disjoint. The association rule $X \rightarrow Y$ holds in a transaction set D with *confidence c*, if $c\%$ of the transactions in D that contain X, also contain Y. The association rule $X \rightarrow Y$ has *support s* in transaction set D, if $s\%$ of the transactions in D contain $X \cup Y$. From these definitions, we see that confidence corresponds to the strength of a rule, while support corresponds to statistical significance. Those rules which exceed a predetermined minimum threshold for support and confidence are considered to be interesting. Syntactic constraints can also be used to restrict the items that can appear in the left- or right-hand side of a rule [2, 131].

2.2.4. Rule Templates

Rule templates [82] are an extension of the syntactic constraints described in [2] and are used to describe a pattern for those attributes that can appear in the left- or right-hand side of an association rule. A *rule template* is given by

$$A_1, A_2, \ldots, A_k \to A_m,$$

where each A_i is either an attribute name, a class name (a class hierarchy is used to map database values to a taxonomy of classes), or an expression $C+$ or $C*$. In the expressions $C+$ and $C*$, C is a class name and $C+$ and $C*$ correspond to one or more, and zero or more, instances of the class C, respectively. An induced rule matches the pattern specified in a rule template if it can be considered to be an instance of the pattern. Rule templates may be either inclusive or restrictive. An inclusive rule template specifies desirable rules that are considered to be interesting, while a restrictive rule template specifies undesirable rules that are considered to be uninteresting. Rule pruning can be done by setting support, confidence, and rule size thresholds.

2.2.5. Projected Savings

In some domains involving financial applications, interesting deviations from some normative or expected values are those that indicate the existence of a problem that can be corrected by some relevant action that will produce financial benefit or payback. Projected savings [95] is a measure that assesses the financial impact of such a benefit or payback. *Projected savings* is given by

$$PS = PI * SP,$$

where PI is the projected impact and SP is the savings percentage. The *projected impact* is given by

$$PI = PD * PF,$$

where PD is the difference between the current average cost and the normative or expected cost for some product or service, and PF is the impact factor (which may be viewed, say, as the number of units of some product or service sold). The savings percentage, SP, is a domain expert specified value of the percentage decrease in deviation that would result following some relevant intervention strategy. The interestingness of a deviation is directly related to the projected savings achievable as a result of this strategy.

2.2.6.   *I*-Measures

The *I*-measures [49] are used to quantify the significance of discovered knowledge, presented in the form of generalized relations or summaries, based upon the structure of concept hiercharchies associated with the attributes in the original ungeneralized relation. The I_1 measure considers the number of non-ANY, non-leaf nodes in a summary, and is given by

$$I_1 = \sum_v c(t(v)),$$

where v is an attribute value, $t(v)$ is the concept hierarchy associated with the attribute containing v, and $c(t(v))$ is a function that returns 1 if v is a non-ANY, non-leaf concept, and 0 otherwise. The I_2 measure considers the depth and weighted height for all nodes in a summary, and is given by

$$I_2 = \sum_v (k)d(v, t(v)) + (1 - k)wh(v, t(v)),$$

where k specifies the relative significance of the depth of a concept versus the weighted depth (e.g., $k = 0.5$ indicates the distance from the root node is as significant as the distance from the leaf nodes), v is an attribute value, $t(v)$ is the concept hierarchy associated with the attribute containing v, $d(v, t(v))$ is the depth of v in concept hierarchy $t(v)$, and $wh(v, t(v))$ is the weighted height of v in concept hierarchy $t(v)$. The depth $d(v, t(v))$ of v in concept hierarchy $t(v)$ is defined so that the depth of the root node is zero and the depth of any other concept is one more than the depth of its parent. The weighted height $wh(v, t(v))$ of v in concept hierarchy $t(v)$ is a function of the number of leaf concepts it has as descendants and the sum of the distances to each of its descendants. Summaries associated with higher values of I_1 and I_2 are considered more interesting.

2.2.7.   Silbershatz and Tuzhilin's Interestingness

Interestingness [126] determines the extent to which a soft belief is changed as a result of encountering new evidence (i.e., discovered knowledge). A soft belief is one that a user is willing to change as new evidence is encountered. *Interestingness*, within the context of soft beliefs, is given by

$$I = \sum_\alpha \frac{|p(\alpha|E, \varepsilon) - p(\alpha|\varepsilon)|}{p(\alpha|\varepsilon)},$$

where α is a belief, E is new evidence, ε is the previous evidence supporting belief α, $p(\alpha|\varepsilon)$ is the confidence in belief α, and $p(\alpha|E, \varepsilon)$ is the new confidence in belief α given the new evidence E. Summation is over all beliefs.

Bayes Theorem is used to determine the new confidence, and is given by

$$p(\alpha|E, \varepsilon) = \frac{p(E|\alpha, \varepsilon)p(\alpha|\varepsilon)}{p(E|\alpha, \varepsilon)p(\alpha|\varepsilon) + p(E|\neg\alpha, \varepsilon)p(\neg\alpha|\varepsilon)}.$$

Positive (negative) evidence strengthens (weakens) the belief.

2.2.8. Kamber and Shinghal's Interestingness

Interestingness [79] is used to determine the relative rank of classification rules based upon necessity and sufficiency. There are two kinds of classification rules: discriminant and characteristic. A *discriminant rule*, $e \rightarrow h$, where e is evidence and h is a hypothesis, summarizes the conditions sufficient to distinguish one class from another. *Sufficiency* is given by

$$S(e \rightarrow h) = \frac{p(e|h)}{p(e|\neg h)}.$$

A *characteristic rule*, $h \rightarrow e$, summaries the conditions necessary for membership in a class. *Necessity* is given by

$$N(e \rightarrow h) = \frac{p(\neg e|h)}{p(\neg e|\neg h)}.$$

Necessity and sufficiency can be used to assess the interestingness of the characteristic rule $h \rightarrow e$, as follows

$$IC^{++} = \begin{cases} (1 - N(e \rightarrow h)) \times p(h), 0 \le N(e \rightarrow h) < 1 \\ 0, \text{otherwise} \end{cases},$$

$$IC^{+-} = \begin{cases} (1 - S(e \rightarrow h)) \times p(h), 0 \le S(e \rightarrow h) < 1 \\ 0, \text{otherwise} \end{cases},$$

$$IC^{-+} = \begin{cases} (1 - 1/N(e \rightarrow h)) \times p(\neg h), 1 < N(e \rightarrow h) < \infty \\ 0, \text{otherwise} \end{cases},$$

and

$$IC^{--} = \begin{cases} (1 - 1/S(e \rightarrow h)) \times p(\neg h), 1 < S(e \rightarrow h) < \infty \\ 0, \text{otherwise} \end{cases},$$

where $++$, $+-$, $-+$, and $--$ correspond to the characteristic rules $h \rightarrow e$, $h \rightarrow \neg e$, $\neg h \rightarrow e$, and $\neg h \rightarrow \neg e$, respectively. Interestingness values for each of the measures lies in $[0, 1]$, where 0 and 1 represent the minimum and maximum possible interestingness, respectively.

2.2.9.   Credibility

Credibility [52] determines the extent to which a classification (i.e., generalized relation or summary) provides decisions for all or nearly all possible values of the condition attributes, based upon adequately supported evidence. *Credibility* is given by

$$Cred_E(C) = Q_E(C) \times min(I/M, 1),$$

where E is an equivalence class, C is a classification, $Q_E(C)$ is the quality of classification C, I is the actual number of instances supporting the equivalence class E, M is the minimum number of instances required for a credible classification, and $min(I/M, 1)$ is a factor that ensures a proportional weight is associated to equivalence classes not supported by an adequate number of instances. The *quality function*, $Q_E(C)$, is given by

$$Q_E(C) = \beta \times p(E) \times |p(F|E) - p(F)|,$$

where β is a normalization factor to ensure that $Q_E(C)$ is always within the interval $[0, 1]$, $P(E)$ is the probability of equivalence class E, $P(F|E)$ is the conditional probability of the occurrence of the concept F (i.e., the decision attribute)   given   that E

has occurred, and $P(F)$ is the probability of concept F. The *normalization factor* is given by

$$\beta = \frac{1}{2p(F)(1 - p(F))}.$$

2.2.10.   General Impressions

A general impression [88] is used to evaluate the importance of classification rules by comparing discovered rules to an approximate or vague description of what is considered to be interesting. So, a general impression is a kind of specification language. There are two types of general impressions that can be specified: Type 1 and Type 2. A *Type 1 general impression* is a rule of the form $A_1OP_1, A_2OP_2, \ldots, A_xOP_x \rightarrow C_j$, where each A_iOP_i is called an impression term, each A_i is an attribute, each OP_i is an impression descriptor from the set $\{<, >, \ll, |, []\}$, and C_j is a class. The $<$ ($>$) impression descriptor means smaller (larger) attribute values are more likely to lead to inclusion in class C_j, $\ll$ means some range of attribute values are more likely to lead to inclusion in class C_j, $|$ means some relationship exists between an attribute and class C_j but the nature of this relationship is not exactly known, and $[]$ means that some subset of the possible values for an attribute are more likely to lead to inclusion in class C_j. A Type 2 general impression is specified when there is more confidence that the combination of impression terms will lead to inclusion in class C_j. A *Type 2 general impression* is a rule of the form $A_1 O P_1, A_2 O P_2, \ldots, A_k O P_k \& A_m O P_m, A_n O P_n, \ldots, A_x O P_x \rightarrow C_j$, where the part to the left (right) of the $\&$ symbol is called the core (supplement). The core must always exist, otherwise the general impression should be specified as Type 1. If the supplement exists, then the rule is called a maximal impression. In a maximal impression, the general impression is that the impression terms in the core and any subset of those in the supplement are more likely to lead to inclusion in class C_j. If the supplement does not exist, then the rule is called an exact impression. In an exact impression, the general impression is that the impression terms in the core are more likely to lead to inclusion in class C_j. The specified general impressions are matched against the rules generated, and ranked to identify those that are most valid.

2.2.11. Distance Metric

The distance metric [39] measures the distance between two rules and is used to determine the rule that provides the highest coverage for the given data. The *distance metric* is given by

$$D(R_i, R_j) = \begin{cases} (\frac{DA(R_i,R_j)+2DV(R_i,R_j)-2EV(R_i,R_j)}{N(R_i)+N(R_j)}, NO(R_i, R_j) = 0 \\ 2, \text{otherwise} \end{cases},$$

where R_i and R_j are rule i and j, respectively, $DA(R_i, R_j)$ is the sum of the number of attributes in R_i not in R_j and the number of attributes in R_j not in R_i, $DV(R_i, R_j)$ is the number of attributes in R_i and R_j that have slightly overlapping values in the range conditions (slightly overlapping means less than 66% of the range), $EV(R_i, R_j)$ is the number of attributes in R_i and R_j that have overlapping values in the range conditions (overlapping means more than 66% of the range), $N(R_i)$ and $N(R_j)$ are the number of attributes in R_i and R_j, respectively, and $NO(R_i, R_j)$ is the number of attributes in R_i and R_j with nonoverlapping values. The distance metric returns a value on $[-1, 1]$ or 2. Values near -1 and 1 indicate a strong and slight overlap, respectively, while the value 2 indicates no overlap. The rule with the highest average distance to the other rules is considered to be the most interesting.

2.2.12. Surprisingness

Surprisingness [38] is a measure that determines the interestingness of discovered knowledge via the explicit detection of occurrences of Simpson's paradox. Simpson's paradox is described, as follows. Let X_1 and X_2 be two mutually exclusive and exhaustive populations partitioned according to the value of a binary event attribute E, where E_1 and E_2 are the values of E in X_1 and X_2, respectively. Let $P(E_1)$ and $P(E_2)$ be the probabilities of events E_1 and E_2 in X_1 and X_2, respectively. Now let X_1 and X_2 be further partitioned according to the value of a categorical attribute having m distinct values (i.e., event E_i is partitioned into events E_{ij}, $i = 1, 2$ and $j = 1, 2, \ldots, m$). Then let $P(E_{1j})$ and $P(E_{2j})$ be the probabilities for events E_{1j} and E_{2j} in X_1 and X_2, respectively. Assuming that $P(E_1) > P(E_2)$ ($P(E_1) < P(E_2)$), Simpson's paradox occurs when $P(E_{1j}) \leq P(E_{2j})$ ($P(E_{1j}) \geq P(E_{2j})$) for all $j = 1, 2, \ldots, m$. That is, although event E_1 (E_2) has a higher (lower) overall probability of occurring, the probability of occurrence of each E_{1j} (E_{2j}) in E_1 (E_2) can actually be lower (higher) than each E_{2j} (E_{1j}) in E_2 (E_1).

2.2.13. Gray and Orlowska's Interestingness

Interestingness [44] is used to evaluate the strength of associations between sets of items in retail transactions (i.e., association rules). While support and confidence have been shown to be useful for characterizing association rules, interestingness contains a discrimination component that gives an indication of the independence of the antecedent and consequent. *Interestingness* is given by

$$I = \left(\left(\frac{P(X \cap Y)}{P(X) \times P(Y)} \right)^k - 1 \right) \times (P(X) \times P(Y))^m,$$

where $P(X \cap Y)$ is the confidence, $P(X) \times P(Y)$ is the support, $\frac{P(X \cap Y)}{P(X) \times P(Y)}$ is the discrimination, and k and m are parameters to weight the relative importance of the discrimination and support components, respectively. Higher values of interestingness are considered more interesting.

2.2.14. Dong and Li's Interestingness

Interestingness [30] is used to evaluate the importance of an association rule by considering its unexpectedness in terms of other association rules in its neighborhood. The neighborhood of an association rule consists of all association rules within a given distance. The *distance metric* is given by

$$D(R_1, R_2) = \delta_1 \times |(X_1 \cup Y_1)\Theta(X_2 \cup Y_2)| + \delta_2 \times |X_1 \Theta X_2| + \delta_3 \times |Y_1 \Theta Y_2|,$$

where $R_1 = X_1 \to Y_1$, $R_2 = X_2 \to Y_2$, δ_1, δ_2, and δ_3 are parameters to weight the relative importance of all three terms, and Θ is an operator denoting the symmetric difference between X and Y (i.e., $(X - Y) \cup (Y - X)$). An *r-neighborhood* of a rule is given by the set

$$N(R_0, r) = \{R | D(R, R_0) \leq r, \text{R a potential rule}\},$$

and is used to define the interestingness of a rule. Two types of interestingness are: unexpected confidence and isolated. *Unexpected confidence interestingness* is given by

$$UCI = \begin{cases} 1, \text{if } ||c(R_0) - ac(R_0, r)| - sc(R_0, r)| > t_1 \\ 0, \text{otherwise} \end{cases},$$

where $c(R_0)$ is th confidence of R_0, $ac(R_0, r)$ and $sc(R_0, r)$ are the average confidence and standard deviation of the rules in the set $M \cap N(R_0, r) - \{R_0\}$ (M is the set of rules satisfying the minimum support and confidence), and t_1 is a threshold. *Isolated interestingness* is given by

$$II = \left\{ \begin{array}{l} 1, \text{if } |N(R_0, r)| - |M \cap N(R_0, r)| > t_2 \\ 0, \text{otherwise} \end{array} \right. ,$$

where $|N(R_0, r)|$ is the number of potential rules in an r-neighborhood, $|M \cap N(R_0, r)|$ is the number of rules generated from the neighborhood, and t_2 is a threshold.

2.2.15. Reliable Exceptions

A reliable exception [89] is a weak rule having relatively small support and relatively high confidence. Reliable exceptions can be induced, as follows. First, use rule induction to generate the strong rules (or some predetermined number of the strongest rules according to some threshold). Reliable exceptions will be evaluated with respect to these strong rules. Second, using contingency table analysis, identify significant deviations between the actual and expected frequency of occurrence of attribute-value and class pairs. Third, specify a deviation threshold. For positive (negative) deviations, any deviation greater than (less than) the threshold is considered outstanding. Fourth, get all instances containing the attribute-value and class pairs of the outstanding negative deviations (i.e., all instances satisfying the rule $X \rightarrow c$, where X is an attribute-value and c is a class. Fifth, calculate the difference between the confidence of the rule $X \rightarrow c$ for the selected instances and the whole dataset. Now the confidence for the selected instances is always 1. So, a large difference (i.e., near 1) implies that the confidence on the whole dataset is low (i.e., near 0), and thus, a reliable exception has been discovered.

2.2.16. Peculiarity

Peculiarity [145] is used to determine the extent to which one data object differs from other similar data objects. *Peculiarity* is given by

$$P(x_i) = \sum_{j=1}^{n} \sqrt{N(x_i, x_j)},$$

where x_i and x_j are attribute values, n is the number of different attribute values, and $N(x_i, x_j)$ is the conceptual distance between x_i and x_j. The *conceptual difference* is given by

$$N(x_i, x_j) = |x_i - x_j|.$$

Chapter 3

A DATA MINING TECHNIQUE

The data mining step in KDD specifies the task to be performed, such as summarization or anomaly-detection. In this chapter, we introduce the data structures and algorithms utilized by our data mining technique. These data structures and algorithms have been incorporated into DGG-Discover and DGG-Interest, extensions to DB-Discover, a research software tool for KDD developed at the University of Regina [22, 23, 24, 25, 50, 65, 66, 70, 72].

3.1. Definitions

Given a set $S = \{s_1, s_2, \ldots, s_n\}$ representing the domain of some attribute, we can partition S in many different ways. For example, $D_1 = \{\{s_1\}, \{s_2\}, \ldots, \{s_n\}\}$, $D_2 = \{\{s_1\}, \{s_2, \ldots, s_n\}\}$, and $D_3 = \{\{s_1, s_2\}, \{s_3, \ldots, s_n\}\}$ represent three possible partitions of S. Let $D = \{D_1, D_2, \ldots, D_m\}$ be the set of partitions of set S. We define a nonempty binary relation $\preceq$ (called a *generalization relation*) on D, where we say $D_i \preceq D_j$ if for every $d_i \in D_i$ there exists $d_j \in D_j$ such that $d_i \subseteq d_j$. The generalization relation $\preceq$ is a partial order relation and $\langle D, \preceq \rangle$ defines a partial order set from which we can construct a graph called a *domain generalization graph*, or DGG, $\langle D, E \rangle$ as follows. First, the nodes of the graph are elements of D. And second, there is a directed edge from D_i to D_j (denoted by $E(D_i, D_j)$) iff $D_i \neq D_j$, $D_i \preceq D_j$, and there is no $D_k \in D$ such that $D_i \preceq D_k$ and $D_k \preceq D_j$.

Let $D_g = \{S\}$ and $D_d = \{\{s_1\}, \{s_2\}, \ldots, \{s_n\}\}$. For any $D_i \in D$ we have $D_d \preceq D_i$ and $D_i \preceq D_g$, where D_d and D_g are called the *least* and *greatest elements* of D, respectively. We call the nodes (elements of D) *domains*, where the least element is the *most specific level of generality* and the greatest element is the *most general level*. A DGG may exist where the least element is mapped directly to the greatest element (i.e., D_d is mapped to D_g).

For example, given $S = \{1, 2, 3, 4, 5, 6, 7\}$ let $D = \{$*Office, City, Division, Country, ANY*$\}$, where $D_g = \{S\} = \{\{1, 2, 3, 4, 5, 6, 7\}\}$, $D_3 = \{\{1, 2\}, \{3, 4, 5, 6, 7\}\}$, $D_2 = \{\{1, 2, 3, 4\}, \{5, 6, 7\}\}$, $D_1 = \{\{1, 2\}, \{3, 4\}, \{5, 6, 7\}\}$, and $D_d = \{\{1\}, \{2\}, \{3\}, \{4\}, \{5\}, \{6\}, \{7\}\}$. If the generalization relation associated with the edge from the *City* node to the *Country* node in the DGG of Figure 1.2 is defined as $\{$*Vancouver* $\rightarrow$ *Canada, Los Angeles* $\rightarrow$ *United States, New York* $\rightarrow$ *United States*$\}$, then the partitions D_d, D_1, D_2, D_3, and D_g correspond to the nodes in the multi-path DGG shown in Figure 1.2, where D_d, D_1, D_2, D_3, and D_g correspond to the *Office, City, Division, Country,* and *ANY* nodes, respectively.

Finally, let a *summary* S be a relation defined on the columns $\{(A_1, D_1)$, $(A_2, D_2), \ldots, (A_n, D_n)\}$, where each (A_i, D_i) is an attribute-domain pair. Also, let $\{(A_1, v_{i1}), (A_2, v_{i2}), \ldots, (A_n, v_{in})\}$, $i = 1, 2, \ldots, m$, be a set of m unique tuples, where each (A_j, v_{ij}) is an attribute-value pair and each v_{ij} is a value from the domain D_j associated with attribute A_j. One attribute A_k is a derived attribute, called *Count*, whose domain D_k is the set of positive integers, and whose value v_{ik} for each attribute-value pair (A_k, v_{ik}) is equal to the number of tuples which have been aggregated from the base relation (i.e., the unconditioned data present in the original relational database). For example, the summary shown in Table 1.2 conforms to this definition.

3.2.   The Serial Algorithm

Given a relation R, a set of m DGGs, and a set of m attributes, where one DGG is associated with each attribute, the *All_Gen* algorithm, shown in Figure 3.1, generates all possible summaries consistent with the DGGs for the set of attributes. In *All_Gen*, the function Node_Count (line 4) determines the number of nodes in DGG D_i. The function Generalize (line 9) returns a summary where attribute i in the input relation has been generalized to the level of node D_{i_k} (that is, D_{i_k} is the k-th node of DGG D_i). Any of the generalization algorithms presented in [22, 23, 24, 25, 54] may be used to implement the Generalize function. The procedure Interest (line 10) determines the interestingness of the summary. The procedure Output (line 11) saves the summary and combination of nodes from which the summary was generated.

3.2.1.   General Overview

The initial call to *All_Gen* is All_Gen $(R, 1, m, D, D_{nodes})$, where R is the input relation for the discovery task, 1 is an identifier corresponding to the first attribute, m is an identifier corresponding to the last attribute, D is the set of m DGGs associated with the m attributes, and D_{nodes} is a vector in which the i-th element is initialized to D_{i_1} (we assume the first node in each D_i corresponds

```
1       procedure All_Gen (relation, i, m, D, Dnodes)
2       begin
3           work_relation ← relation
4           for k = 1 to Node_Count (Di) − 1 do begin
5               if i < m then
6                   All_Gen (work_relation, i + 1, m, D, Dnodes)
7               end
8               Dnodes[i] ← Di k+1
9               work_relation ← Generalize (relation, i, Di k+1)
10              interestingness ← Interest (work_relation)
11              Output (work_relation, Dnodes, interestingness)
12          end
13      end
```

Figure 3.1.   Serial multi-attribute generalization algorithm

to the domain of D_i). D_{nodes} is used to store the combination of nodes from which each summary is generated.

The algorithm is described as follows. In the i-th iteration of *All_Gen* (corresponding to the i-th attribute), one pass is made through the *for* loop (lines 4 to 11) for each non-domain node in D_i (i.e., the DGG associated with attribute i). If the i-th iteration of *All_Gen* is not also the m-th iteration (that is, corresponding to the last attribute) (line 5), then the $i + 1$-th iteration of *All_Gen* is made (line 6). The $i + 1$-th iteration of *All_Gen* is All_Gen ($work\_relation, i + 1, m, D, D_{nodes}$), where the values of m, D, and D_{nodes} do not change from the i-th iteration. The first parameter, $work\_relation$, was previously set to the value of $relation$ prior to entering the for loop (line 3). The second parameter, i, is incremented by one (corresponding to the $i + 1$-th attribute). In the first pass through the for loop (i.e., $k = 1$) for the i-th iteration, the value of $work\_relation$ is R (i.e., the original input relation).

In the m-th iteration of *All_Gen*, or when the $i + 1$-th iteration returns control to the i-th iteration (line 6), the i-th iteration determines the next level of generalization for attribute i (i.e., $D_{i_{k+1}}$) and saves it in the i-th element of the vector D_{nodes} (line 8). The relation used as input to the i-th iteration of *All_Gen* is generalized to the level of node $D_{i_{k+1}}$ (line 9), the interestingness of the resulting summary is determined (line 10), and the summary is saved along with the interestingness and combination of nodes from which the summary was generated (line 11). In all passes through the for loop, other than the first (i.e., $k > 1$), the value of $work\_relation$ passed by the i-th iteration to the $i + 1$-th iteration is the value of $relation$ generalized to the level of $D_{i_{k+1}}$.

3.2.2. Detailed Walkthrough

We now present a detailed walkthrough of the serial algorithm. Consider again the sales database shown in Table 1.1, and the associated DGGs for the *Shape, Size,* and *Colour* attributes shown in Figure 1.3. The *All_Gen* procedure is initially called with parameters *relation* = the contents of the sales database from Table 1.1, $i = 1$, $m = 3$, D = the DGGs from Figure 1.3, and $D_{nodes} = \{\langle Shape \rangle, \langle Size \rangle, \langle Colour \rangle\}$. In this walkthrough, we assume that $D_{1_1} = \langle Shape \rangle$, $D_{1_2} = \langle ANY \rangle$, $D_{2_1} = \langle Size \rangle$, $D_{2_2} = \langle Package \rangle$, $D_{2_3} = \langle Weight \rangle$, $D_{2_4} = \langle ANY \rangle$, $D_{3_1} = \langle Colour \rangle$, and $D_{3_2} = \langle ANY \rangle$. The initial values of D_{2_2} and D_{2_3} could be swapped (i.e., $D_{2_2} = \langle Weight \rangle$ and $D_{2_3} = \langle Package \rangle$) as the order of their values is arbitrary and of no consequence to the algorithm.

Iteration 1 - Invocation 1 - Loop 1. We set *work_relation* to *relation* (line 3) to prevent changing the original value of *relation* in this iteration. Since $k = 1 \leq$ Node_Count$(D_1) - 1 = 1$ (line 4), we continue with the first loop of this iteration. Since $i = 1 < m = 3$ (line 5), we call the second iteration of *All_Gen* (line 6) with parameters *work_relation* (unchanged from the current iteration), $i = 2$, $m = 3$, D (D never changes from one iteration to the next), and $D_{nodes} = \{\langle Shape \rangle, \langle Size \rangle, \langle Colour \rangle\}$.

Iteration 2 - Invocation 1 - Loop 1. We set *work_relation* to *relation* (line 3). Since $k = 1 \leq$ Node_Count$(D_2) - 1 - 3$ (line 4), we continue with the first loop of this iteration. Since $i = 2 < m = 3$ (line 5), we call the third iteration of *All_Gen* (line 6) with parameters *work_relation* (unchanged from the current iteration), $i = 3$, $m = 3$, D, and $D_{nodes} = \{\langle Shape \rangle, \langle Size \rangle, \langle Colour \rangle\}$.

Iteration 3 - Invocation 1 - Loop 1. We set *work_relation* to *relation* (line 3). Since $k = 1 \leq$ Node_Count$(D_3) - 1 = 1$ (line 4), we continue with the first loop of this iteration. Since $i = 3 \not< m = 3$ (line 5), we do not call a fourth iteration of *All_Gen* (line 6). Instead, we set $D_{nodes}[3] = D_{3_2}$ (i.e., $\langle ANY \rangle$) (line 8), so $D_{nodes} = \{\langle Shape \rangle, \langle Size \rangle, \langle ANY \rangle\}$. We set *work_relation* to the result returned from a call to Generalize (line 9) with parameters *relation*, $i = 3$, and D_{3_2}. The value of *work_relation*, shown in Table 3.1, is the value of Table 1.1, having selected only the *Shape, Size,* and *Colour* attributes, with the *Colour* attribute generalized to the level of node D_{3_2}.

We call Interest (line 10) with parameter *work_relation*. We call Output (line 11) with parameters *work_relation* and $D_{nodes} = \{\langle Shape \rangle, \langle Size \rangle, \langle ANY \rangle\}$. Since $k =$ Node_Count$(D_3) - 1 = 1$ (line 4), the first invocation of the third iteration is complete.

Table 3.1. Summary for the DGG node combination *Shape-Size-ANY*

Shape	Size	Colour	Count
round	small	ANY	2
square	medium	ANY	1
round	large	ANY	2
square	x-large	ANY	1
square	small	ANY	1

Iteration 2 - Invocation 1 - Loop 1 (continued). The call to the third iteration of *All_Gen* (line 6) is complete. We set $D_{nodes}[2] = D_{2_2}$ (i.e., $\langle Package \rangle$) (line 8), so $D_{nodes} = \{\langle Shape \rangle, \langle Package \rangle, \langle Colour \rangle\}$. We set *work_relation* to the result returned from a call to Generalize (line 9) with parameters *relation*, $i = 2$, and D_{2_2}. The value of *work_relation*, shown in Table 3.2, is the value of Table 1.1, having selected only the *Shape, Size,* and *Colour* attributes, with the *Size* attribute generalized to the level of node D_{2_2}. We call Interest (line 10) with parameter *work_relation*. We call Output (line 11) with parameters *work_relation* and $D_{nodes} = \{\langle Shape \rangle, \langle Package \rangle, \langle Colour \rangle\}$. The first loop of the first invocation of the second iteration is complete.

Table 3.2. Summary for the DGG node combination *Shape-Package-Colour*

Shape	Size	Colour	Count
round	bag	white	2
square	bag	black	2
round	box	white	1
round	box	black	1
square	box	white	1

Iteration 2 - Invocation 1 - Loop 2. Since $k = 2 \leq$ Node_Count(D_2) $-$ $1 = 3$ (line 4), we continue with the second loop of this iteration. Since $i = 2 < m = 3$ (line 5), we call the third iteration of *All_Gen* (line 6) with parameters *work_relation* (unchanged the current iteration), $i = 3, m = 3, D,$ and $D_{nodes} = \{\langle Shape \rangle, \langle Package \rangle, \langle Colour \rangle\}$.

Iteration 3 - Invocation 1 - Loop 1. We set *work_relation* to *relation* (line 3). Since $k = 1 \leq$ Node_Count(D_3) $- 1 = 1$ (line 4), we continue with the first loop of this iteration. Since $i = 3 \not< m = 3$ (line 5), we do not call a fourth iteration of *All_Gen* (line 6). Instead, we set $D_{nodes}[3] = D_{3_2}$ (i.e., $\langle ANY \rangle$) (line 8), so $D_{nodes} = \{\langle Shape \rangle, \langle Package \rangle, \langle ANY \rangle\}$. We set

Table 3.3.   Summary for the DGG node combination *Shape-Package-ANY*

Shape	Size	Colour	Count
round	bag	ANY	2
square	bag	ANY	2
round	box	ANY	2
square	box	ANY	1

work_relation to the result returned from a call to Generalize (line 9) with parameters *relation*, $i = 3$, and D_{3_2}. The value of *work_relation*, shown in Table 3.3, is the value of Table 1.1, having selected only the *Shape*, *Size*, and *Colour* attributes, with the *Colour* attribute generalized to the level of node D_{3_2}. We call Interest (line 10) with parameter *work_relation*. We call Output (line 11) with parameters *work_relation* and $D_{nodes} = \{ \langle Shape \rangle, \langle Package \rangle, \langle ANY \rangle \}$. Since $k = $ Node_Count$(D_3) - 1 = 1$ (line 4), the second invocation of the third iteration is complete.

The important aspects of the serial algorithm have now been clearly demonstrated.

3.3. The Parallel Algorithm

As previously mentioned in Section 1.1.1, the size of the generalization space depends only on the number of nodes in the DGGs; it is not dependent upon the number of tuples in the input relation or the number of attributes selected for generalization. When the number of attributes to be generalized is large, or the DGGs associated with a set of attributes are complex, meaning the generalization space is large, we can improve the performance of the serial algorithm through parallel generalization. However, our parallel algorithm does not simply assign one node in the generalization space to each processor, because the startup cost for each processor was found to be too great in comparison to the actual work performed. Through experimentation, we adopted a more coarsegrained approach, where a unique combination of paths, consisting of one path through the DGG associated with each attribute, is assigned to each processor. For example, given attribute A with three possible paths through its DGG, attribute B with 4, and attribute C with 2, our approach creates $3 \times 4 \times 2 = 24$ processes. The *Par_All_Gen* algorithm, shown in Figure 3.2, creates parallel *All_Gen* child processes on multiple processors (line 8). In *Par_All_Gen*, the function Path_Count (line 3) determines the number of paths in DGG D_i.

```
1        procedure Par_All_Gen (relation, i, m, D, D_paths, D_nodes)
2        begin
3            for k = 1 to Path_Count (D_i) do begin
4                D_paths[i] ← D_i^k
5                if i < m then
6                    Par_All_Gen (relation, i + 1, m, D, D_paths, D_nodes)
7                else
8                    fork All_Gen (relation, 1, m, D_paths, D_nodes)
9                end
10           end
11       end
```

Figure 3.2.   Parallel multi-attribute generalization algorithm

3.3.1.   General Overview

The initial call to *Par_All_Gen* is Par_All_Gen $(R, 1, m, D, \emptyset, D_{nodes})$, where $R, 1, m, D$, and D_{nodes} have the same meaning as in the serial algorithm, and $\emptyset$ initializes D_{paths}. D_{paths} is a vector in which the i-th element is assigned a unique path from D_i.

The algorithm is described as follows. In the i-th iteration of *Par_All_Gen*, one pass is made through the *for* loop (lines 3 to 10) for each distinct path in D_i. The current path, D_i^k, is determined and saved in the k-th element of D_{paths} (line 4), where D_i^k is the k-th path in D_i. If the i-th iteration of *Par_All_Gen* is not also the m-th iteration (line 5), then the $i+1$-th iteration of *Par_All_Gen* is called (line 6). The $i + 1$-th iteration of *Par_All_Gen* is Par_All_Gen $(relation, i + 1, m, D, D_{paths}, D_{nodes})$, where the values for $relation, m, D$, and D_{nodes} do not change from the i-th iteration. The second parameter is incremented by one. The fifth parameter, D_{paths}, was previously set to D_i^k (line 4). When the $i + 1$-th iteration returns control to the i-th iteration (line 6), the next pass through the *for* loop begins (line 4).

In the m-th iteration of *Par_All_Gen*, an *All_Gen* child process is created (line 8). The call to All_Gen is All_Gen $(relation, 1, m, D_{paths}, D_{nodes})$, where $relation, m$, and D_{nodes} are unchanged from the values passed as parameters to the m-th iteration of *Par_All_Gen*. The second parameter, 1, is an identifier corresponding to the first attribute. The fourth parameter, D_{paths}, is a unique vector containing m paths (i.e., one from each D_i associated with the set of attributes). The All_Gen child process then follows the serial algorithm described in the previous section.

The parallel algorithm may generalize the same combination of nodes in D_{nodes} on multiple processors. This can occur when a node in a DGG resides on more than one path (i.e., the node is at an intersection where two or more paths cross). This could be prevented through prior analysis of the generalization space or by introducing some form of communication and synchronization between processors, but both approaches introduce additional overhead. For these experiments, we consider this redundant generalization to be tolerable because it only occurs in a small percentage of the total number of states in the generalization space. However, in general, the number of redundant states may certainly be domain specific and, in some circumstances, could be a limiting factor in some applications of the proposed algorithm.

3.3.2.   Detailed Walkthrough

We now present a detailed walkthrough of the parallel algorithm. Consider again the sales database shown in Table 1.1, and the associated DGGs for the *Shape*, *Size*, and *Colour* attributes shown in Figure 1.3. The *Par_All_Gen* procedure is initially called with parameters *relation* = the contents of the sales database from Table 1.1, $i = 1$, $m = 3$, D = the DGGs from Figure 1.3, $D_{paths} = \emptyset$, and $D_{nodes} = \{\langle Shape \rangle, \langle Size \rangle, \langle Colour \rangle\}$. We assume that the D_{i_j} have the same values as described in Section 3.2.2. In this walkthrough, we also assume that D_1^1 = the path $\langle Shape, \text{ANY} \rangle$ in the DGG for the *Shape* attribute, D_2^1 = the path $\langle Size, Package, \text{ANY} \rangle$ and D_2^2 = the path $\langle Size, Weight, \text{ANY} \rangle$ in the DGG for the *Size* attribute, and D_3^1 = the path $\langle Colour, \text{ANY} \rangle$ in the DGG for the *Colour* attribute.

Iteration 1 - Invocation 1 - Loop 1. Since $k = 1 \leq \text{Path\_Count}(D_1) = 1$ (line 3), we continue with the first loop of this iteration. We set $D_{paths}[1] = D_1^1$ (line 4), so $D_{paths} = \{\langle D_1^1 \rangle\}$. Since $i = 1 < m = 3$ (line 5), we call the second iteration of *Par_All_Gen* (line 6) with parameters *relation* (*relation* never changes from one iteration to the next), $i = 2$, $m = 3$, D (D never changes from one iteration to the next), $D_{paths} = \{\langle D_1^1 \rangle\}$, and $D_{nodes} = \{\langle Shape \rangle, \langle Size \rangle, \langle Colour \rangle\}$.

Iteration 2 - Invocation 1 - Loop 1. Since $k = 1 \leq \text{Path\_Count}(D_2) = 2$ (line 3), we continue with the first loop of this iteration. We set $D_{paths}[2] = D_2^1$ (line 4), so $D_{paths} = \{\langle D_1^1 \rangle, \langle D_2^1 \rangle\}$. Since $i = 2 < m = 3$ (line 5), we call the third iteration of *Par_All_Gen* (line 6) with parameters *relation*, $i = 3$, $m = 3$, D, $D_{paths} = \{\langle D_1^1 \rangle, \langle D_2^1 \rangle\}$, and $D_{nodes} = \{\langle Shape \rangle, \langle Size \rangle, \langle Colour \rangle\}$.

Iteration 3 - Invocation 1 - Loop 1. Since $k = 1 \leq \text{Path\_Count}(D_3) = 1$ (line 3), we continue with the first loop of this iteration. We set $D_{paths}[3] = D_3^1$ (line 4), so $D_{paths} = \{\langle D_1^1 \rangle, \langle D_2^1 \rangle, \langle D_3^1 \rangle\}$. Since $i = 3 \not< m = 3$ (line 5), we do not call a fourth iteration of All_Gen (line 6). Instead, we fork a call to All_Gen (line 8) with parameters *relation*, $i = 1$, $m = 3$, $D = D_{paths}$, and $D_{nodes} = \{\langle Shape \rangle, \langle Package \rangle, \langle Colour \rangle\}$. All_Gen will execute the serial algorithm, as described in Section 3.2.2, to generate all summaries described by the nodes on the paths in D. Since $k = $ Path_Count$(D_3) = 1$ (line 3), the first invocation of the third iteration is complete.

Iteration 2 - Invocation 1 - Loop 1 (continued). The call to the third iteration of *Par_All_Gen* (line 6) is complete. The first loop of the first invocation of the second iteration is complete.

Iteration 2 - Invocation 1 - Loop 2. Since $k = 2 \leq$ Path_Count$(D_2) = 2$ (line 3), we continue with the second loop of this iteration. We set $D_{paths}[2] = D_2^2$ (line 4), so $D_{paths} = \{\langle D_1^1 \rangle, \langle D_2^2 \rangle\}$. Since $i = 2 < m = 3$ (line 5), we call the third iteration of *Par_All_Gen* (line 6) with parameters *relation*, $i = 3$, $m = 3$, D, $D_{paths} = \{\langle D_1^1 \rangle, \langle D_2^2 \rangle\}$, and $D_{nodes} = \{\langle Shape \rangle, \langle Size \rangle, \langle Colour \rangle\}$.

Iteration 3 - Invocation 2 - Loop 1. Since $k = 1 \leq$ Path_Count$(D_3) = 1$ (line 3), we continue with the first loop of this iteration. We set $D_{paths}[3] = D_3^1$ (line 4), so $D_{paths} = \{\langle D_1^1 \rangle, \langle D_2^2 \rangle, \langle D_3^1 \rangle\}$. Since $i = 3 \not< m = 3$ (line 5), we do not call a fourth iteration of All_Gen (line 6). Instead, we fork a call to All_Gen (line 8) with parameters *relation*, $i = 1$, $m = 3$, $D = D_{paths}$, and $D_{nodes} = \{\langle Shape \rangle, \langle Package \rangle, \langle Colour \rangle\}$. Since $k = $ Path_Count$(D_3) = 1$ (line 3), the second invocation of the third iteration is complete.

Iteration 2 - Invocation 1 - Loop 2 (continued). The call to the third iteration of *Par_All_Gen* (line 6) is complete. Since $k = $ Path_Count$(D_2) = 2$ (line 3), the first invocation of the second iteration is complete.

The important aspects of the parallel algorithm have now been clearly demonstrated.

3.4. Complexity Analysis

3.4.1. Attribute-Oriented Generalization

Attribute-oriented generalization is a summarization technique that has been effective for KDD. As the result of recent research, AOG methods are considered among the most efficient of KDD methods for knowledge discovery from databases [22, 23, 24, 54, 77]. In particular, algorithms for generating summaries from relational databases are presented in [23, 25] that run in $O(n)$ time, where n is the number of tuples in the input relation, and require $O(p)$ space, where p is the number of tuples in the summaries (typically $p << n$). In [25], it is also proven that an AOG algorithm which runs in $O(n)$ time is optimal for generalizing a relation.

3.4.2. The All_Gen Algorithm

In general, for a discovery task containing m attributes, a database containing n tuples, and an $O(n)$ generalization algorithm, creating all possible summaries requires $O(n \prod_{i=1}^{m} |D_i|)$ time, where $|D_i|$ is the number of nodes in the DGG for attribute i.

3.5.   A Comparison with Commercial OLAP Systems

Currently available OLAP tools, such as BrioQuery from Brio Technologies, Crystal Reports from Seagate Software, and PowerPlay from Cognos have the capability to perform generalization at consolidation time. Here we compare our multi-attribute generalization technique to that of PowerPlay (we believe PowerPlay to be representative of commercial OLAP tools). The generalization technique utilized in PowerPlay is similar to that described in [21, 24, 25, 56].

In PowerPlay, a *dimension* is analogous to a single-path DGG. A collection of dimensions is referred to as a *dimension map* and describes the universe (i.e., generalization space) for a particular set of attributes. A sample dimension map is shown in Table 3.4, where the *Dim 1*, *Dim 2*, and *Dim 3* columns correspond to the ⟨ *Shape*, ANY ⟩, ⟨ *Size*, *Package*, ANY ⟩, and ⟨ *Colour*, ANY ⟩ paths of the DGG shown in Figure 1.3. One dimension is associated with each attribute. A discovery task is referred to as an *application*, and is limited to an exploration of the universe defined by the dimension map. Multiple dimensions can be associated with an attribute (similar to a multi-path DGG associated with an attribute), but each dimension associated with the attribute requires that an additional column be added to the application. Alternatively, separate applications can be defined, each containing an alternative dimension for the attribute. Both methods require the user to manage the relationships between dimensions when defining the applications and when analyzing the results. Summaries at particular levels of aggregation/generalization must be preselected by the user, and interesting summaries must be located by manually traversing the universe. If an application is not supplying satisfactory results, the user can switch to another application and re-query the database. For any but small applications, the actual size of the universe, the number of possible summaries, and what each may contain is usually not known. Rules of thumb suggest limiting the number of dimensions to no more than five and the number of levels in each dimension to no more than seven. Finally, there are no parallel processing capabilities.

In our approach, we allow multiple DGGs (i.e., dimensions) to be associated with an attribute. We then explicitly generate all possible summaries in the generalization space, and rank the interestingness of each summary before presenting the results to the user. Interesting summaries can then be used as a starting point for further exploration and analysis. If a group of related summaries (i.e., associated with a particular combination of DGG paths) is not supplying satisfactory results, the user can jump to another group of related summaries (i.e., associated with a different combination of DGG paths) without requiring the user to re-query the database. A description of the level of

Table 3.4. A sample dimension map for the *Shape*, *Size*, and *Colour* attributes

Level	Dim 1	Dim 2	Dim 3
1	All Shapes	All Sizes	All Colours
	1	1	1
2	Shape	Package	Colour
	2	2	2
3		Size	
		4	

generalization/aggregation for each attribute is provided with each summary. We also suggest a rule of thumb where the number of attributes is limited to no more than five, but we do not suggest any limits on the number of levels in the associated DGGs. Although both techniques allow more than five attributes/dimensions, interpretation of results becomes more difficult. For large problems, where many attributes have been selected for generalization, or the DGGs associated with the attributes have many nodes, we provide a parallel technique which partitions the problem across multiple processors.

Chapter 4

HEURISTIC MEASURES OF INTERESTINGNESS

The tuples in a generalized relation (i.e., a summary generated from a database) are unique, and therefore, can be considered to be a population with a structure that can be described by some frequency or probability distribution based upon the values contained in the derived *Count* attribute. In this chapter, we describe sixteen diversity measures that evaluate the frequency or probability distribution of the values in the derived *Count* attribute in a summary to assign a single real-valued index that represents its interestingness relative to other summaries generated from the same database. The measures are well-known measures of dispersion, dominance, inequality, and concentration that have previously been successfully and frequently applied in several areas of the physical, social, ecological, management, information, and computer sciences. Their use for ranking summaries generated from databases is a new application area.

4.1. Diversity

Diversity is an important concept that has seen extensive use in several different areas of research. However, although diversity is used in many disparate areas, it is widely claimed that diversity is a difficult concept to define [8, 9, 74, 76, 92, 108, 109, 137]. The difficulty in defining diversity arises because it actually encompasses two separate components: the number of *classes* (also referred to in the literature as *richness*, *abundance*, or *density*) and the *proportional distribution* of the population among the classes (also referred to in the literature as *relative abundance*, *heterogeneity*, or *evenness*). Within the context of ranking the interestingness of a summary, the number of classes is simply the number of tuples in the summary; the proportional distribution is simply the actual probability distribution of the classes based upon the values contained in the derived *Count* attribute.

In a typical diversity measure, the two components are combined to characterize the variability of a population by a single value. This concept of a dual-component diversity measure was first introduced in [127], and strongly supported in [74, 76, 138]. The diversity measures considered to be most useful, and those most frequently referenced in the literature, are dual-component measures. Yet, despite the widespread acceptance and use of diversity measures, there is no single mathematical definition of diversity which has been widely accepted as the de facto standard and which has been shown to be superior to all others [8, 92, 108]. There is some general agreement, however, that a population is considered to have high diversity when it has many classes and the proportional distribution is fairly even. Similarly, a population is considered to have low diversity when it has few classes and the proportional distribution is uneven. Unfortunately, this leaves considerable room for ambiguity in measuring diversity because a population with few classes and a fairly even proportional distribution could have the same or nearly the same diversity as a population with many classes and an uneven proportional distribution.

Although there are some problems related to a precise and universally accepted definition for diversity, there are numerous research areas where the concept of diversity has been considered useful. In ecology, various measures of diversity have been proposed and studied to aid in understanding the variability of populations of organisms within different types of habitat [8, 16, 20, 96, 102, 109]. Diversity measures have also been used by economists and social scientists to study the distribution of income between different socioeconomic groups or geographical regions [9, 12, 28, 123]. As an aid to understanding genetic differences between populations, diversity measures have been used by geneticists and biologists [85]. Diversity measures have been used to describe the linguistic differences between the inhabitants of neighboring geographic regions [46, 86]. In business, diversity measures have been useful for determining the extent of industrial concentration or market penetration of consumer products [13, 61, 62, 75]. Other applications of diversity measures have occurred in the areas of epidemiology [78], bibliometrics [111], software engineering [107], and the measurement of scientific productivity [9]. More general treatments attempt to define the concept of diversity and develop a related theory of diversity measurement [108, 137].

Here we apply sixteen diversity measures to a new application area, that of ranking the interestingness of summaries generated from databases. They share three important properties. First, each measure depends only on the frequency or probability distribution of the values in the derived *Count* attribute of the summary to which it is being applied. Second, each measure allows a value to be generated with at most one pass through the summary. And third, each measure is independent of any specific units of measure. Utilizing these heuristics for ranking the interestingness of summaries generated from databases is a natural and useful extension for these diversity measures into a new application domain.

4.2. Notation

Variables used to describe the interestingness measures in the presentation that follows are now defined. Let m be the total number of tuples in a summary. Let n_i be the value contained in the derived *Count* attribute for tuple t_i. Let $N = \sum_{i=1}^{m} n_i$ be the total count. Let p be the actual probability distribution of the tuples based upon the values n_i. Let $p_i = n_i/N$ be the actual probability for tuple t_i. Let q be a uniform probability distribution of the tuples. Let $\bar{u} = N/m$ be the count for tuple t_i, $i = 1, 2, \ldots, m$ according to the uniform distribution q. Let $\bar{q} = 1/m$ be the probability for tuple t_i, for all $i = 1, 2, \ldots, m$ according to the uniform distribution q. Let r be the probability distribution obtained by combining the values n_i and $\bar{u}$. Let $r_i = (n_i + \bar{u})/2N$, be the probability for tuples t_i, for all $i = 1, 2, \ldots, m$ according to the distribution r. For example, given the sample summary shown in Table 4.1, we have $m = 4$, $n_1 = 3$, $n_2 = 2$, $n_3 = 1$, $n_4 = 1$, $N = 7$, $p_1 = 0.429$, $p_2 = 0.286$, $p_3 = 0.143$, $p_4 = 0.143$, $\bar{u} = 1.75$, $\bar{q} = 0.25$, $r_1 = 0.339$, $r_2 = 0.268$, $r_3 = 0.196$, and $r_4 = 0.196$.

Table 4.1. A sample summary

Colour	Shape	Count
red	round	3
green	round	2
red	square	1
blue	square	1

4.3. The Sixteen Diversity Measures

We now describe sixteen diversity measures and provide a detailed example of the calculation of each measure. Each example provided is based upon the sample summary previously shown in Table 4.1.

4.3.1. The $I_{Variance}$ Measure

The $I_{Variance}$ measure, based upon sample variance from classical statistics [120], measures the weighted average of the squared deviations of the probabilities p_i from the mean probability $\bar{q}$, where the weight assigned to each squared deviation is $1/(m-1)$. We use sample variance because we assume the summary may not contain all possible combinations of attribute values, meaning we are not observing all of the possible tuples. The *sample variance* is given by

$$I_{Variance} = \frac{\sum_{i=1}^{m}(p_i - \bar{q})^2}{m-1}.$$

Example 4.1. The calculation of the $I_{Variance}$ measure is illustrated below.

$$
\begin{aligned}
I_{Variance} &= ((0.429 - 0.25)^2 + (0.286 - 0.25)^2 + (0.143 - 0.25)^2 \\
&\quad + (0.143 - 0.25)^2)/3 \\
&= 0.018
\end{aligned}
$$

4.3.2.   The $I_{Simpson}$ Measure

The $I_{Simpson}$ measure, a variance-like measure based upon the Simpson index [127], measures the extent to which the counts are distributed over the tuples in a summary, rather than being concentrated in any single one of them. The *concentration* is given by

$$I_{Simpson} = \sum_{i=1}^{m} p_i^2.$$

Background. Let each tuple i be represented by a "commonness value" (i.e., the probability of occurrence p_i). If an individual is drawn at random from the population, the probability that it will belong to tuple i is p_i, and if it does, its commonness value is also p_i. Thus, the expected commonness values for tuple i is p_i^2, and for all tuples $i = 1, \ldots, n$ is $\sum_{1}^{m} p_i^2$. Equivalently, this can be viewed as the average commonness value that would be obtained if the experiment of drawing an individual at random were repeated many times.

Example 4.2. The calculation of the $I_{Simpson}$ measure is illustrated below.

$$
\begin{aligned}
I_{Simpson} &= 0.429^2 + 0.286^2 + 0.143^2 + 0.143^2 \\
&= 0.306
\end{aligned}
$$

4.3.3.   The $I_{Shannon}$ Measure

The $I_{Shannon}$ measure, based upon a relative entropy measure from information theory (known as the *Shannon index*) [125], measures the average information content in the tuples of a summary. The *average information content*, in bits per tuple, is given by

$$I_{Shannon} = -\sum_{i=1}^{m} p_i \log_2 p_i.$$

Background. Say there are n_i individuals summarized in a tuple i, out of a possible N individuals. The probability of drawing one of the individuals in tuple i is n_i/N, or p_i. The information conveyed by announcing the result of drawing a particular individual in tuple i is $-\log_2 p_i$. The total contribution of these n_i individuals to the overall average information conveyed by announcing the result is $-p_i \log_2 p_i$. Summation over all such cases for all possible individuals is given by $-\sum_{i=1}^{m} p_i \log_2 p_i$.

Example 4.3. The calculation of the $I_{Shannon}$ measure is illustrated below.

$$
\begin{aligned}
I_{Shannon} &= -(0.429 \log_2 0.429 + 0.286 \log_2 0.286 + 0.143 \log_2 0.143 \\
&\quad +0.143 \log_2 0.143) \\
&= 1.842 \; bits
\end{aligned}
$$

4.3.4.   The I_{Total} Measure

The I_{Total} measure, also based upon the Shannon index from information theory [139], measures the total information content in a summary. The *total information content*, in bits, is given by

$$ I_{Total} = m * I_{Shannon}. $$

Example 4.4. The calculation of the I_{Total} measure is illustrated below.

$$
\begin{aligned}
I_{Total} &= 4 * 1.842 \\
&= 7.368 \; bits
\end{aligned}
$$

4.3.5.   The I_{Max} Measure

The I_{Max} measure, also based upon the Shannon index from information theory [139], measures the maximum possible information content in a summary. The *maximum possible information content*, in bits, is given by

$$ I_{Max} = \log_2 m. $$

Background. The maximum information per tuple occurs when each of the tuples has equal probability. For example, consider a summary containing four tuples where $p_i = 0.25$ for all $i = 1, \ldots, 4$. Then

$$
\begin{aligned}
I_{Max} &= 4(0.25 \log_2 0.25) \\
&= 4(0.25(2)) \\
&= 4(0.5) \\
&= 2 \; bits
\end{aligned}
$$

Example 4.5. The calculation of the I_{Max} measure is illustrated below.

$$
\begin{aligned}
I_{Max} &= \log_2 4 \\
&= 2 \; bits
\end{aligned}
$$

4.3.6.   The $I_{McIntosh}$ Measure

The $I_{McIntosh}$ measure, based upon a heterogeneity index from ecology [96], views the counts in a summary as the coordinates of a point in a multidimensional space and measures the modified Euclidean distance from this point to the origin. The *modified Euclidean distance* is given by

$$I_{McIntosh} = \frac{N - \sqrt{\sum_{i=1}^{m} n_i^2}}{N - \sqrt{N}}.$$

Background. The value $\sqrt{\sum_{i=1}^{m} n_i^2}$ is just the Pythagorean Theorem. Since $\sqrt{\sum_{i=1}^{m} n_i^2}$ is a measure of concentration, the N-complement $N - \sqrt{\sum_{i=1}^{m} n_i^2}$ is a measure of diversity. The value $N - \sqrt{N}$ makes it a diversity measure independent of N. The greater the count in a particular class, the further that class will be from the origin. If the count is reduced, or the count is spread more evenly between class, the distance from the origin will be reduced. $I_{McIntosh}$ relates the distance between a class and the origin to the range of possible values as determined by the number of tuples in the original relation.

Example 4.6. The calculation of the $I_{McIntosh}$ measure is illustrated below.

$$
\begin{aligned}
I_{McIntosh} &= \frac{7 - \sqrt{3^2 + 2^2 + 1^2 + 1^2}}{7 - \sqrt{7}} \\
&= 0.718
\end{aligned}
$$

4.3.7.   The I_{Lorenz} Measure

The I_{Lorenz} measure, based upon the Lorenz curve from statistics, economics, and social science [133], measures the average value of the Lorenz curve derived from the probabilities p_i associated with the tuples in a summary. The *average value of the Lorenz curve* is given by

$$I_{Lorenz} = \bar{q} \sum_{i=1}^{m} (m - i + 1) p_i.$$

Background. The Lorenz curve is a series of straight lines in a square of unit length, starting from the origin and going successively to points (p_1, q_1), $(p_1 + p_2, q_1 + q_2)$, . . ., as shown in Figure 4.1. When the p_i's are all equal, the Lorenz curve coincides with the diagonal that cuts the unit square into equal halves. When the p_i's are not all equal, the Lorenz curve is below the diagonal.

Example 4.7. The calculation of the I_{Lorenz} measure is illustrated below.

$$I_{Lorenz} = (0.25)(4p_4 + 3p_3 + 2p_2 + p_1)$$

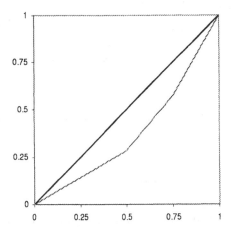

Figure 4.1. A sample Lorenz curve

$$
\begin{aligned}
&= \quad (0.25)((4)(0.143)+(3)(0.143)+(2)(0.286)+(1)(0.429)) \\
&= \quad 0.501
\end{aligned}
$$

4.3.8. The I_{Gini} Measure

The I_{Gini} measure, based upon the Gini coefficient [133], which is itself defined in terms of the Lorenz curve, measures the ratio of the area between the diagonal (i.e., the line of equality) and the Lorenz curve, and the total area below the diagonal. The *Gini coefficient* is given by

$$
I_{Gini} = \frac{\bar{q}\sum_{i=1}^{m}\sum_{j=1}^{m}|p_i - p_j|}{2}.
$$

Example 4.8. The calculation of the I_{Gini} measure is illustrated below.

$$
\begin{aligned}
I_{Gini} = \quad &(0.25)((|0.429 - 0.429| + |0.429 - 0.143| \\
&+|0.429 - 0.143| + |0.429 - 0.286|) \\
&+(|0.286 - 0.429| + |0.286 - 0.143| \\
&+|0.286 - 0.143| + |0.286 - 0.286|) \\
&+(|0.143 - 0.429| + |0.143 - 0.143| \\
&+|0.143 - 0.143| + |0.143 - 0.286|) \\
&+(|0.143 - 0.429| + |0.143 - 0.143| \\
&+|0.143 - 0.143| + |0.143 - 0.286|))/2 \\
= \quad &0.252
\end{aligned}
$$

4.3.9.   The I_{Berger} Measure

The I_{Berger} measure, based upon a dominance index from ecology [14], measures the proportional dominance of the tuple in a summary with the highest probability p_i. The *proportional dominance* is given by

$$I_{Berger} = \max(p_i).$$

Example 4.9. The calculation of the I_{Berger} measure is illustrated below.

$$I_{Berger} \quad = \quad 0.429$$

4.3.10.   The I_{Schutz} Measure

The I_{Schutz} measure, based upon an inequality measure from economics and social science [123], measures the relative mean deviation of the actual distribution of the counts in a summary from a uniform distribution of the counts. The *relative mean deviation* is given by

$$I_{Schutz} = \frac{\sum_{i=1}^{m} |p_i - \bar{q}|}{2m\bar{q}}.$$

Example 4.10. The calculation of the I_{Schutz} measure is illustrated below.

$$
\begin{aligned}
I_{Schutz} \quad = \quad & (|0.429 - 0.25| + |0.286 - 0.25| + |0.143 - 0.25| \\
& + |0.143 - 0.25|)/(2)(4)(0.25) \\
= \quad & 0.215
\end{aligned}
$$

4.3.11.   The I_{Bray} Measure

The I_{Bray} measure, based upon a community similarity index from ecology [16], measures the percentage of similarity between the actual distribution of the counts in a summary and a uniform distribution of the counts. The *percentage of similarity* is given by

$$I_{Bray} = \frac{\sum_{i=1}^{m} \min(n_i, \bar{u})}{N}.$$

Example 4.11. The calculation of the I_{Bray} measure is illustrated below.

$$
\begin{aligned}
I_{Bray} \quad &= \quad \frac{1.75 + 1.75 + 1.0 + 1.0}{7} \\
&= \quad 0.786
\end{aligned}
$$

4.3.12.   The $I_{Whittaker}$ Measure

The $I_{Whittaker}$ measure, based upon a community similarity index from ecology [138], measures the percentage of similarity between the actual distribution of the counts in a summary and a uniform distribution of the counts.

The *percentage of similarity* is given by

$$I_{Whittaker} = 1 - \left(\frac{\sum_{i=1}^{m} |p_i - \bar{q}|}{2} \right).$$

Example 4.12. The calculation of the $I_{Whittaker}$ measure is illustrated below.

$$
\begin{aligned}
I_{Whittaker} &= 1 - (|0.429 - 0.25| + |0.286 - 0.25| + |0.143 - 0.25| \\
&\quad + |0.143 - 0.25|)/2 \\
&= 0.786
\end{aligned}
$$

4.3.13. The $I_{Kullback}$ Measure

The $I_{Kullback}$ measure, based upon a distance measure and the $I_{Shannon}$ measure from information theory [84], measures the distance between the actual distribution of the counts in a summary and a uniform distribution of the counts. The *distance*, in bits, is given by

$$I_{Kullback} = \log_2 m - \sum_{i=1}^{m} p_i \log_2 \frac{p_i}{\bar{q}}.$$

Example 4.13. The calculation of the $I_{Kullback}$ measure is illustrated below.

$$
\begin{aligned}
I_{Kullback} &= 2 - \left(0.429 \log_2 \frac{0.429}{0.25} + 0.286 \log_2 \frac{0.286}{0.25} \right. \\
&\quad \left. + 0.143 \log_2 \frac{0.143}{0.25} + 0.143 \log_2 \frac{0.143}{0.25} \right) \\
&= 1.842 \ bits
\end{aligned}
$$

4.3.14. The $I_{MacArthur}$ Measure

The $I_{MacArthur}$ measure, based upon the Shannon index from information theory [91], combines two summaries and then measures the difference between the amount of information contained in the combined distribution and the amount contained in the average of the two original distributions. The *difference*, in bits, is given by

$$I_{MacArthur} = \left(-\sum_{i=1}^{m} r_i \log_2 r_i \right) - \left(\frac{(-\sum_{i=1}^{m} p_i \log_2 p_i) + \log_2 m}{2} \right).$$

Example 4.14. The calculation of the $I_{MacArthur}$ measure is illustrated below.

$$
\begin{aligned}
I_{MacArthur} &= -(0.339 \log_2 0.339 + 0.268 \log_2 0.268 + 0.196 \log_2 0.196 \\
&\quad + 0.196 \log_2 0.196) - (-(0.429 \log_2 0.429 \\
&\quad + 0.286 \log_2 0.286 + 0.143 \log_2 0.143 + 0.143 \log_2 0.143) \\
&\quad - \log_2 4)/2) \\
&= 0.039 \ bits
\end{aligned}
$$

4.3.15.   The I_{Theil} Measure

The I_{Theil} measure, based upon a distance measure from information theory [133], measures the distance between the actual distribution of the counts in a summary and a uniform distribution of the counts. The *distance*, in bits, is given by

$$I_{Theil} = \frac{\sum_{i=1}^{m} |p_i \log_2 p_i - \bar{q} \log_2 \bar{q}|}{m\bar{q}}.$$

Example 4.15. The calculation of the I_{Theil} measure is illustrated below.

$$
\begin{aligned}
I_{Theil} &= (|0.429 \log_2 0.429 - 0.25 \log_2 0.25| + |0.286 \log_2 0.286 \\
&\quad -0.25 \log_2 0.25| + |0.143 \log_2 0.143 - 0.25 \log_2 0.25| \\
&\quad +|0.143 \log_2 0.143 - 0.25 \log_2 0.25|)/(4)(0.25) \\
&= 0.238 \; bits
\end{aligned}
$$

4.3.16.   The $I_{Atkinson}$ Measure

The $I_{Atkinson}$ measure, based upon a measure of inequality from economics [12], measures the percentage to which the population in a summary would have to be increased to achieve the same level of interestingness if the counts in the summary were uniformly distributed. The *percentage increase* is given by

$$I_{Atkinson} = 1 - \left(\prod_{i=1}^{m} \frac{p_i}{\bar{q}} \right)^{\bar{q}}.$$

Background. Lower values of $I_{Atkinson}$ mean that the distribution of counts in a summary are fairly equal, or near uniform. Higher values mean the distribution is fairly uneven. As an example, say $I_{Atkinson} = 0.105$, as shown in the example below. This value means that if the counts of the tuples were uniformly distributed, then we would need only approximately 90% of the current total count to realize the same level of interestingness.

Example 4.16. The calculation of the $I_{Atkinson}$ measure is illustrated below.

$$
\begin{aligned}
I_{Atkinson} &= 1 - \left(\frac{0.429}{0.25} * \frac{0.286}{0.25} * \frac{0.143}{0.25} * \frac{0.143}{0.25} \right)^{0.25} \\
&= 0.105
\end{aligned}
$$

Chapter 5

AN INTERESTINGNESS FRAMEWORK

In this chapter, we develop a theory of interestingness for the ranking of summaries generated from databases. The theory provides the foundation for an intuitive understanding of the term "interestingness" when used within this context.

5.1.   Interestingness Principles

In this section, we develop a theory of interestingness through the mathematical formulation of five independent principles that must be satisfied by any acceptable measure of interestingness for ranking summaries generated from databases. We study functions f of m variables, $f(n_1, \ldots, n_m)$ (thus f denotes a general interestingness measure), where m and each n_i (n_i assumed to be nonzero) are as defined in Section 4.2, and $(n_1, \ldots, n_m)$ is a vector corresponding to the values in the derived *Count* attribute for some arbitrary summary. For all measures, except I_{Lorenz}, the values in the vector $(n_1, \ldots, n_m)$ are not assumed to be arranged in any particular order. I_{Lorenz} assumes the values are in ascending order so that $n_m \geq \ldots \geq n_1$. We begin by specifying two fundamental principles.

Minimum Value Principle (P1). Given a vector $(n_1, \ldots, n_m)$, where $n_i = n_j$, $i \neq j$, for all i, j, $f(n_1, \ldots, n_m)$ attains its minimum value.

P1 specifies that the minimum interestingness should be attained when the tuple counts are all equal (i.e., uniformly distributed). Proofs for measures satisfying this principle need to show that the uniform distribution generates the lowest possible index value.

Maximum Value Principle (P2). Given a vector $(n_1, \ldots, n_m)$, where $n_1 = N - m + 1$, $n_i = 1$, $i = 2, \ldots, m$, and $N > m$, $f(n_1, \ldots, n_m)$ attains its maximum value.

P2 specifies that the maximum interestingness should be attained when the tuple counts are distributed as unevenly as possible. Proofs for measures satisfying this principle need to show that the most uneven distribution generates the highest possible index value.

The behaviour of a measure relative to satisfying both P1 and P2 is significant because it reveals an important characteristic about its fundamental nature as a measure of diversity. A measure of diversity can generally be considered either a *measure of concentration* or a *measure of dispersion*. A measure of concentration can be viewed as the opposite of a measure of dispersion, and we can convert one to the other via simple transformations. For example, if g corresponds to a measure of dispersion, then we can convert it to a measure of concentration f, where $f = \max(g) - g$. Here we only consider measures of concentration (i.e., those satisfying P1 and P2), so measures of dispersion (i.e., those not satisfying P1 and P2 in a particular way, as discussed in the next section) were transformed into measures of concentration prior to our analysis.

Skewness Principle (P3). Given a vector $(n_1, \ldots, n_m)$, where $n_1 = N - m + 1$, $n_i = 1, i = 2, \ldots, m$, and $N > m$, and a vector $(n_1 - c, n_2, \ldots, n_m, n_{m+1}, \ldots, n_{m+c})$, where $n_1 - c > 1$ and $n_i = 1$, $i = 2, \ldots, m + c$, $f(n_1, \ldots, n_m) > f(n_1 - c, n_2, \ldots, n_m, n_{m+1}, \ldots, n_{m+c})$.

P3 specifies that a summary containing m tuples, whose counts are distributed as unevenly as possible, will be more interesting than a summary containing $m + c$ tuples, whose counts are also distributed as unevenly as possible. Proofs for measures satisfying this principle need to show that a short, maximally uneven distribution generates a higher value than a long, maximally uneven distribution.

Permutation Invariance Principle (P4). Given a vector $(n_1, \ldots, n_m)$ and any permutation $(i_1, \ldots, i_m)$ of $(1, \ldots, m)$, $f(n_1, \ldots, n_m) = f(n_{i_1}, \ldots, n_{i_m})$.

P4 specifies that every permutation of a given distribution of tuple counts should be equally interesting. That is, interestingness is not a labeled property, it is only determined by the distribution of the counts.

Transfer Principle (P5). Given a vector $(n_1, \ldots, n_m)$ and $0 < c < n_j$, $f(n_1, \ldots, n_i + c, \ldots, n_j - c, \ldots, n_m) > f(n_1, \ldots, n_i, \ldots, n_j, \ldots, n_m)$.

P5, adapted from [28], specifies that when a strictly positive transfer is made from the count of one tuple to another tuple whose count is greater, then interestingness increases.

5.2. Summary

In this section, we address the problem of identifying those measures that satisfy the proposed principles. Summarized results are shown in Tables 5.1 and 5.2. In Tables 5.1 and 5.2, the *P1* to *P5* columns describe those measures that satisfy the proposed principles P1 to P5. The *Concentration Order* and *Dispersion Order* columns in Table 5.1 describe those measures that satisfy the proposed principles when all of the measures are considered to be measures of concentration and measures of dispersion, respectively (to be discussed in detail below). The *Aggregate Order* columns in Table 5.2 are an aggregation of the *Concentration Order* and *Dispersion Order* columns, and describe those measures of concentration and transformed measures of dispersion that satisfy the proposed principles. In Tables 5.1 and 5.2, symbols are defined as follows. A measure that has been mathematically proven to satisfy a principle is indicated by the *bullet* symbol (i.e., •). A measure that we conjectured to satisfy a principle is indicated by the *circle* symbol (i.e., ○). A measure that satisfies a principle in both the *Concentration Order* and *Dispersion Order* columns of Table 5.1 is indicated by the *plus* symbol (i.e., +) in the *Aggregate Order* columns of Table 5.2. A measure that fails to satisfy a principle in either of the *Concentration Order* or *Dispersion Order* columns of Table 5.1 is indicated by the *times* symbol (i.e., ×) in the *Aggregate Order* columns of Table 5.2.

Table 5.1. Measures satisfying the principles (concentration order and dispersion order)

Measure	Concentration Order					Dispersion Order				
	P1	P2	P3	P4	P5	P1	P2	P3	P4	P5
$I_{Variance}$	•	○	•	•	•				•	
$I_{Simpson}$	•	•	•	•	•				•	
$I_{Shannon}$			•			•	•	•	•	•
I_{Total}			•			•	•	•	•	•
I_{Max}			•					•	•	
$I_{McIntosh}$			•			•	•	•	•	•
I_{Lorenz}			•			•	•			•
I_{Gini}	•	•		•	•			•	•	
I_{Berger}	•	•	•	•					•	
I_{Schutz}	•	•		•				•	•	
I_{Bray}			•	•		•	•		•	
$I_{Whittaker}$			•	•		•	•		•	
$I_{Kullback}$			•			•	•	•	•	•
$I_{MacArthur}$	•	•		•	•			•	•	
I_{Theil}	•		•					•	•	
$I_{Atkinson}$	•	•		•	•			•	•	

Table 5.2.　Measures satisfying the principles (aggregate order)

Measure	Aggregate Order				
	P1	P2	P3	P4	P5
$I_{Variance}$	●	○	●	+	●
$I_{Simpson}$	●	●	●	+	●
$I_{Shannon}$	●	●	●	+	●
I_{Total}	●	●	●	+	●
I_{Max}	×	×	●	+	×
$I_{McIntosh}$	●	●	●	+	●
I_{Lorenz}	●	●		×	●
I_{Gini}	●	●		+	●
I_{Berger}	●	●	●	+	×
I_{Schutz}	●	●		+	×
I_{Bray}	●	●		+	×
$I_{Whittaker}$	●	●		+	×
$I_{Kullback}$	●	●	●	+	●
$I_{MacArthur}$	●	●		+	●
I_{Theil}	●	×		+	×
$I_{Atkinson}$	●	●		+	●

In the *Concentration Order* columns of Table 5.1, higher index values generated by a measure are considered to be more interesting. This corresponds to an ordering of index values where the lowest value is associated with a uniform distribution of the counts (according to P1), and the highest value is associated with the most uneven distribution of the counts (according to P2). That we consider higher index values to be more interesting is just an arbitrary starting point for our analysis and will be shown to be an assumption with no significant consequences. In the *Concentration Order* columns of Table 5.1, we see that $I_{Variance}$ satisfies both P1 and P2. We also see that $I_{Shannon}$ does not satisfy P1 and P2. Similarly, in the *Dispersion Order* columns Table 5.1, lower index values generated by a measures are considered to be more interesting. This corresponds to an ordering of index values where the highest value is associated with a uniform distribution of the counts (contrary to P1), and the lowest value is associated with the most uneven distribution of the counts (contrary to P2). Now in the *Dispersion Order* columns of Table 5.1, we see that $I_{Shannon}$ satisfies both P1 and P2. We also see that $I_{Variance}$ does not satisfy P1 and P2. So, the order in which summaries are ranked by $I_{Shannon}$ is opposite to the order in which they are ranked by $I_{Variance}$ (at least for the summaries represented by a uniform and the most uneven distribution of the counts). Other measures are seen to behave in a similar manner. Within the context of our analysis, a measure that satisfies both P1 and P2, such as

$I_{Variance}$ does in the *Concentration Order* columns of Table 5.1, is considered to be a measure of concentration, while a measure that satisfies both P1 and P2, such as $I_{Shannon}$ does in the *Dispersion Order* columns of Table 5.1, is considered to be a measure of dispersion.

We could have done our analysis by considering only measures of dispersion. In that case, we could have assumed that lower index values in the current *Concentration Order* columns of Table 5.1 were to be considered more interesting. Those measures satisfying the proposed principles would then have been identical to those shown in the current *Dispersion Order* columns of Table 5.1. Similarly, in the *Dispersion Order* columns of Table 5.1, we could have assumed that higher index values were to be considered more interesting. Those measures satisfying the proposed principles would then have been identical to those shown in the current *Concentration Order* columns of Table 5.1. We then would have used a transformation function h, similar to f, to convert selected measures of concentration into measures of dispersion. The end result is that those measures satisfying the proposed principles in the *Aggregate Order* columns of Table 5.2 would have been identical to those currently shown. So the initial assumption that higher values are to be considered more interesting is of no consequence in our analysis. That is, in the proofs that follow, where we show $f(x) > f(y)$ for measures of concentration, we show equivalently that $f(x) < f(y)$ for measures of dispersion.

5.3. Theorems and Proofs

In this section, we state theorems indicating those measures that satisfy the principles in the *Aggregate Order* columns of Table 5.2, and derive proofs for each. We also state theorems for those measures that fail to satisfy particular principles, but leave the proofs as an exercise for the reader (vectors are suggested that will clearly demonstrate the failure). At the beginning of each section, we restate the relevant principle for reader convenience.

5.3.1. Minimum Value Principle

Minimum Value Principle (P1). Given a vector $(n_1, \ldots, n_m)$, where $n_i = n_j$, $i \neq j$, for all i, j, $f(n_1, \ldots, n_m)$ attains its minimum value.

Theorem 5.1. $I_{Variance}$ satisfies P1.

Proof. We need to show that

$$\sum_{i=1}^{m} \left(\frac{n_i}{N} - \frac{1}{m} \right)^2 = 0.$$

The values in the vector $(n_1, \ldots, n_m)$ are uniformly distributed, so $n_1 = \ldots = n_m$, $n_1/N = \ldots = n_m/N$, and all $p_i = n_i/N$ are equal, for all i. Now

$$\bar{q} = \frac{\sum_{i=1}^{m} \left(\frac{n_i}{N} \right)}{m} = \frac{m \left(\frac{n_i}{N} \right)}{m} = \frac{n_i}{N},$$

and

$$\bar{q} = \frac{\sum_{i=1}^{m} \left(\frac{n_i}{N} \right)}{m} = \frac{1}{m},$$

so it is shown that $n_i/N = 1/m$, for all i. Now,

$$\sum_{i=1}^{m} \left(\frac{n_i}{N} - \frac{1}{m} \right)^2 = \sum_{i=1}^{m} \left(\frac{1}{m} - \frac{1}{m} \right)^2 = \sum_{i=1}^{m} 0 = 0,$$

and it is proved.

Theorem 5.2. $I_{Simpson}$ satisfies P1.

Proof. We need to show that when the values in the vector $(n_1, \ldots, n_m)$ are uniformly distributed, $f(n_1, \ldots, n_m) = 1/m$. That is, we need to show that

$$\sum_{i=1}^{m} \left(\frac{n_i}{N} \right)^2 = \frac{1}{m}.$$

We also need to show that for any vector $(n_1 + c, n_2 - d_2, \ldots n_m - d_m)$ (i.e., a vector whose values are not uniformly distributed), where at least one $d_i \neq 0$, $d_i < n_i$, and $\sum_{i=2}^{m} d_i = c$, $f(n_1 + c, n_2 - d_2, \ldots, n_m - d_m) > 1/m$. That is, we need to show that

$$\left(\frac{n_1 + c}{N} \right)^2 + \sum_{i=2}^{m} \left(\frac{n_i - d_i}{N} \right)^2 > \frac{1}{m}.$$

Now, we have $n_i/N = 1/m$, so

$$\sum_{i=1}^{m} p_i^2 = \sum_{i=1}^{m} \left(\frac{n_i}{N} \right)^2 = \sum_{i=1}^{m} \left(\frac{1}{m} \right)^2 = m \left(\frac{1}{m} \right)^2 = m \left(\frac{1}{m^2} \right) = \frac{1}{m},$$

and the first part is shown. For a vector $(n_1 + c, n_2 - d_2, \ldots, n_m - d_m)$, whose values are not uniformly distributed, showing that

$$\left(\frac{n_1 + c}{N} \right)^2 + \sum_{i=2}^{m} \left(\frac{n_i - d_i}{N} \right)^2 > \frac{1}{m}$$

is equivalent to showing that

$$\left(\frac{n_1 + c}{N} \right)^2 + \sum_{i=2}^{m} \left(\frac{n_i - d_i}{N} \right)^2 - \frac{1}{m} > 0.$$

Since the vector $(n_1 + c, n_2 - d_2, \ldots n_m - d_m)$ is not uniformly distributed, n_1/N is greater than the average $1/m$, so let $1/m = \frac{n_1}{N} - k$. Substituting for $1/m$ in the above inequality yields

$$\left(\frac{n_1 + c}{N}\right)^2 + \sum_{i=2}^{m}\left(\frac{n_i - d_i}{N}\right)^2 - \frac{n_1}{N} + k > 0.$$

Expanding the left side of the above inequality, and putting it over the common denominator N^2 yields

$$n_1^2 + 2n_1 c + c^2 + \sum_{i=2}^{m}(n_i - d_i)^2 - n_1 N + kN^2 > 0.$$

Since $N = n_1 + c + \sum_{i=2}^{m}(n_i - d_i)$, substituting for N in the above inequality yields

$$2n_1\sum_{i=2}^{m}d_i - 2\sum_{i=2}^{m}n_i d_i + \left(\sum_{i=2}^{m}d_i\right)^2 + \sum_{i=2}^{m}n_i^2 + \sum_{i=2}^{m}d_i^2 + kn_1^2$$

$$+2n_1 k\sum_{i=2}^{m}n_i - n_1\sum_{i=2}^{m}n_i + k\left(\sum_{i=2}^{m}n_i\right)^2 + k\left(\sum_{i=2}^{m}d_i\right)^2 > 0.$$

Now two of the positive terms in the above inequality are greater than the two negative terms, specifically,

$$2n_1\sum_{i=2}^{m}d_i > 2\sum_{i=2}^{m}n_i d_i$$

and

$$2n_1 k\sum_{i=2}^{m}n_i > n_1\sum_{i=2}^{m}n_i.$$

The five remaining terms are all positive, so their sum is positive, the second part is shown, and it is proved.

Theorem 5.3. $I_{Shannon}$ satisfies P1.

Proof. We need to show that

$$-\sum_{i=1}^{m}\left(\frac{n_i}{N}\right)\log_2\left(\frac{n_i}{N}\right) >$$

$$-\left(\left(\frac{n_1 + c}{N}\right)\log_2\left(\frac{n_1 + c}{N}\right) + \sum_{i=2}^{m}\left(\frac{n_i - d_i}{N}\right)\log_2\left(\frac{n_i - d_i}{N}\right)\right).$$

Multiplying both sides by -1 yields

$$\sum_{i=1}^{m} \left(\frac{n_i}{N}\right) \log_2 \left(\frac{n_i}{N}\right) <$$

$$\left(\frac{n_1+c}{N}\right) \log_2 \left(\frac{n_1}{N}\right) + \sum_{i=2}^{m} \left(\frac{n_i-d_i}{N}\right) \log_2 \left(\frac{n_i-d_i}{N}\right).$$

Now the values in the vector $(n_1, \ldots, n_m)$ are uniformly distributed, so $n_1 = \ldots = n_m$. Substituting n_1 for n_i yields

$$\sum_{i=1}^{m} \left(\frac{n_1}{N}\right) \log_2 \left(\frac{n_1}{N}\right) <$$

$$\left(\frac{n_1+c}{N}\right) \log_2 \left(\frac{n_1}{N}\right) + \sum_{i=2}^{m} \left(\frac{n_1-d_i}{N}\right) \log_2 \left(\frac{n_1-d_i}{N}\right).$$

Factoring $1/N$ out of both sides of the above inequality yields

$$\sum_{i=1}^{m} n_1 \log_2 \left(\frac{n_1}{N}\right) < (n_1 + c) \log_2 \left(\frac{n_1}{N}\right) + \sum_{i=2}^{m} (n_1 - d_i) \log_2 \left(\frac{n_1-d_i}{N}\right).$$

Using the difference property of logarithms to decompose the above inequality yields

$$\sum_{i=1}^{m} n_1 \log_2 n_1 - \sum_{i=1}^{m} n_1 \log_2 N <$$

$$n_1 \log_2(n_1 + c) + c \log_2(n_1 + c) - n_1 \log_2 N - c \log_2 N$$

$$+ \sum_{i=2}^{m} n_1 \log_2(n_1 - d_i) - \sum_{i=2}^{m} d_i \log_2(n_1 - d_i) - \sum_{i=2}^{m} n_1 \log_2 N$$

$$+ \sum_{i=2}^{m} d_i \log_2 N.$$

Canceling terms yields

$$\sum_{i=1}^{m} n_1 \log_2 n_1 < (n_1 + c) \log_2(n_1 + c) + \sum_{i=2}^{m} (n_1 - d_i) \log_2(n_1 - d_i).$$

The right side of the above inequality can be written as

$$(n_1+c) \log_2(n_1+c) + (n_1-d_2) \log_2(n_1-d_2) + \ldots + (n_1-d_m) \log_2(n_1-d_m).$$

Since the function $x \log_2 x$ is a convex function, we have

$$(n_1 + c) \log_2(n_1 + c) + (n_1 - d_2) \log_2(n_1 - d_2) + \ldots$$
$$+ (n_1 - d_m) \log_2(n_1 - d_m) \geq$$
$$(n_1 + c + n_1 - d_2 + \ldots + n_1 - d_m) \log_2(n_1 + c + n_1 - d_2 + \ldots$$
$$+ n_1 - d_m).$$

Now $c = \sum_{i=2}^{m} d_i$, so substituting for the d_i's on the right side of the above inequality yields

$$(n_1 + c + n_1 - d_2 + \ldots + n_1 - d_m) \log_2(n_1 + c + n_1 - d_2 + \ldots$$
$$+ n_1 - d_m) = (mn_1) \log_2(mn_1),$$

or equivalently

$$(n_1 + c + n_1 - d_2 + \ldots + n_1 - d_m) \log_2(n_1 + c + n_1 - d_2 + \ldots$$
$$+ n_1 - d_m) = \sum_{i=1}^{m} n_1 \log_2(mn_1).$$

We now show that

$$\sum_{i=1}^{m} n_1 \log_2 n_1 < \sum_{i=1}^{m} n_1 \log_2(mn_1).$$

Using the addition property of logarithms to decompose the above inequality yields

$$\sum_{i=1}^{m} n_1 \log_2 n_1 < \sum_{i=1}^{m} n_1 \log_2 m + \sum_{i=1}^{m} n_1 \log_2 n_1.$$

Canceling terms yields

$$0 < \sum_{i=1}^{m} n_1 \log_2 m,$$

and it is proved.

Theorem 5.4. I_{Total} satisfies P1.

Proof. We need to show that

$$m \left(- \sum_{i=1}^{m} \left(\frac{n_i}{N} \right) \log_2 \left(\frac{n_i}{N} \right) \right) >$$
$$m \left(- \left(\left(\frac{n_1 + c}{N} \right) \log_2 \left(\frac{n_1 + c}{N} \right) \right. \right.$$
$$\left. \left. + \sum_{i=2}^{m} \left(\frac{n_i - d_i}{N} \right) \log_2 \left(\frac{n_i - d_i}{N} \right) \right) \right).$$

Canceling terms and multiplying both sides of the above inequality by -1 yields

$$\sum_{i=1}^{m} \left(\frac{n_i}{N}\right) \log_2 \left(\frac{n_i}{N}\right) <$$

$$\left(\frac{n_1 + c}{N}\right) \log_2 \left(\frac{n_1}{N}\right) + \sum_{i=2}^{m} \left(\frac{n_i - d_i}{N}\right) \log_2 \left(\frac{n_i - d_i}{N}\right),$$

which is proved in Theorem 5.3.

Theorem 5.5. $I_{McIntosh}$ satisfies P1.

Proof. We need to show that

$$\frac{N - \sqrt{\sum_{i=1}^{m} n_i^2}}{N - \sqrt{N}} > \frac{N - \sqrt{(n_1 + c)^2 + \sum_{i=2}^{m}(n_i - d_i)^2}}{N - \sqrt{N}}.$$

Canceling terms and multiplying both sides of the above inequality by -1 yields

$$\sum_{i=1}^{m} n_i^2 < (n_1 + c)^2 + \sum_{i=2}^{m}(n_i - d_i)^2.$$

The values in the vector $(n_1, \ldots, n_m)$ are uniformly distributed, so $n_1 = \ldots = n_m$. Substituting n_1 for n_i in the above inequality and expanding the terms on the right side yields

$$mn_1^2 < n_1^2 + 2n_1 c + c^2 + \sum_{i=2}^{m}(n_1^2 - 2n_1 d_i + d_i^2).$$

Distributing the summation term on the right side of the above inequality yields

$$mn_1^2 < n_1^2 + 2n_1 c + c^2 + \sum_{i=2}^{m} n_1^2 - \sum_{i=2}^{m} 2n_1 d_i + \sum_{i=2}^{m} d_i^2.$$

Since $c = \sum_{i=2}^{m} d_i$, substituting for c in the above inequality yields

$$mn_1^2 < n_1^2 + \sum_{i=2}^{m} 2n_1 d_i + \left(\sum_{i=2}^{m} d_i\right)^2 + \sum_{i=2}^{m} n_1^2 - \sum_{i=2}^{m} 2n_1 d_i + \sum_{i=2}^{m} d_i^2.$$

Canceling terms yields

$$0 < \left(\sum_{i=2}^{m} d_i\right)^2 + \sum_{i=2}^{m} d_i^2,$$

and it is proved.

Theorem 5.6. I_{Lorenz} satisfies P1.

Proof. We need to show that

$$\frac{1}{m}\sum_{i=1}^{m}(m-i+1)\left(\frac{n_i}{N}\right) > \frac{1}{m}\left(\sum_{i=1}^{m-1}(m-i+1)\frac{n_i-d_i}{N} + \frac{n_m+c}{N}\right).$$

The values in the vector $(n_1,\ldots,n_m)$ are uniformly distributed, so $n_1 = \ldots = n_m$. Substituting n_m for n_i and distributing the summation on the right side yields

$$\frac{1}{m}\sum_{i=1}^{m-1}(m-i+1)\left(\frac{n_m}{N}\right) + \frac{n_m}{N} >$$
$$\frac{1}{m}\left(\sum_{i=1}^{m-1}(m-i+1)\frac{n_m}{N} - \sum_{i=1}^{m-1}(m-i+1)\frac{d_i}{N} + \frac{n_m}{N} + \frac{c}{N}\right).$$

Canceling terms yields

$$0 > -\sum_{i=1}^{m-1}(m-i+1)\frac{d_i}{N} + \frac{c}{N}.$$

Since $c = \sum_{i=1}^{m-1}d_i$, substituting for c in the above inequality and moving $-\sum_{i=1}^{m-1}(m-i+1)\frac{d_i}{N}$ to the left side yields

$$\sum_{i=1}^{m-1}(m-i+1)\frac{d_i}{N} > \sum_{i=1}^{m-1}\frac{d_i}{N}.$$

Factoring $1/N$ out of both sides of the above inequality yields

$$\sum_{i=1}^{m-1}(m-i+1)d_i > \sum_{i=1}^{m-1}d_i,$$

which is obviously true, and it is proved.

Theorem 5.7. I_{Gini} satisfies P1.

Proof. We need to show that

$$\frac{\sum_{i=1}^{m}\sum_{j=1}^{m}\left|\left(\frac{n_i}{N}\right)\left(\frac{1}{m}\right) - \left(\frac{n_j}{N}\right)\left(\frac{1}{m}\right)\right|}{2} = 0.$$

The values in the vector $(n_1,\ldots,n_m)$ are uniformly distributed, so $n_i/N = 1/m$,

$$\frac{\sum_{i=1}^{m} \sum_{j=1}^{m} \left| \left(\frac{n_1}{N}\right)\left(\frac{1}{m}\right) - \left(\frac{n_1}{N}\right)\left(\frac{1}{m}\right)\right|}{2} = \frac{\sum_{i=1}^{m} \sum_{j=1}^{m} |0|}{2} = 0,$$

and it is proved.

Theorem 5.8. I_{Berger} satisfies P1.

Proof. We need to show that when the values in the vector $(n_1, \ldots, n_m)$ are uniformly distributed, $f(n_1, \ldots, n_m) = n_1/N$. That is, we need to show that

$$\max\left(\frac{n_i}{N}\right) = \frac{n_1}{N}.$$

We also need to show that for any vector $(n_1 + c, n_2 - d_2, \ldots n_m - d_m)$ (i.e., a vector whose values are not uniformly distributed), where at least one $d_i \neq 0$, $d_i < n_i$, and $\sum_{i=2}^{m} d_i = c$, $f(n_1 + c, n_2 - d_2, \ldots, n_m - d_m) > n_1/N$. That is, we need to show that

$$\max\left(\frac{n_i}{N}\right) = \frac{n_1 + c}{N} > \frac{n_1}{N}.$$

The values in the vector $(n_1, \ldots, n_m)$ are uniformly distributed, so $n_1 = \ldots = n_m$, $n_1/N = \ldots = n_m/N$, all $p_i = n_i/N$ are equal, for all i, $\max(p_i) = p_1 = n_1/N$, and the first part is shown. We now show that

$$\frac{n_1 + c}{N} > \frac{n_1}{N}.$$

Factoring $1/N$ out of both sides of the above inequality yields

$$n_1 + c > n_1.$$

Canceling n_1 from both sides yields

$$c > 0.$$

Since $c = \sum_{i=2}^{m}$, and at least one $d_i > 0$, the above inequality is obviously true, the second part is shown, and it is proved.

Theorem 5.9. I_{Schutz} satisfies P1.

Proof. We need to show that

$$\frac{\sum_{i=1}^{m} \left|\frac{n_i}{N} - \frac{1}{m}\right|}{2m\left(\frac{1}{m}\right)} = 0.$$

The values in the vector $(n_1, \ldots, n_m)$ are uniformly distributed, so $n_i/N = 1/m$,

$$\frac{\sum_{i=1}^m \left| \frac{n_i}{N} - \frac{1}{m} \right|}{2m \left(\frac{1}{m} \right)} = \frac{\sum_{i=1}^m \left| \frac{1}{m} - \frac{1}{m} \right|}{2} = \frac{\sum_{i=1}^m |0|}{2} = \frac{0}{2} = 0,$$

and it is proved.

Theorem 5.10. I_{Bray} satisfies P1.

Proof. We need to show that

$$\frac{\sum_{i=1}^m min(n_i, \bar{u})}{N} = 1,$$

which is equivalent to showing that

$$\sum_{i=1}^m min(n_i, \bar{u}) = N.$$

The values in the vector $(n_1, \ldots, n_m)$ are uniformly distributed, so each $n_i = \bar{u}$, for all i. Substituting for $\bar{u}$ in the above equality yields

$$\sum_{i=1}^m min(\bar{u}, \bar{u}) = N,$$

which is equivalent to

$$m\bar{u} = N.$$

Now $N = m\bar{u}$, so substituting for N in the above equality yields

$$N = N,$$

and it is proved.

Theorem 5.11. $I_{Whittaker}$ satisfies P1.

Proof. We need to show that

$$1 - \left(0.5 \sum_{i=1}^m \left| \frac{n_i}{N} - \frac{1}{m} \right| \right) >$$

$$1 - \left(0.5 \left(\left| \frac{n_1 + c}{N} - \frac{1}{m} \right| + \sum_{i=2}^m \left| \frac{n_i - d_i}{N} - \frac{1}{m} \right| \right) \right).$$

Canceling terms and multiplying both sides of the above inequality by -1 yields

$$\sum_{i=1}^m \left| \frac{n_i}{N} - \frac{1}{m} \right| < \left| \frac{n_1 + c}{N} - \frac{1}{m} \right| + \sum_{i=2}^m \left| \frac{n_i - d_i}{N} - \frac{1}{m} \right|.$$

The values in the vector $(n_1, \ldots, n_m)$ are uniformly distributed, so $n_i/N = 1/m$. Substituting for n_i/N in the above inequality yields

$$0 < \left| \frac{n_1 + c}{N} - \frac{1}{m} \right| + \sum_{i=2}^{m} \left| \frac{n_i - d_i}{N} - \frac{1}{m} \right|.$$

Since all the terms on the right side of the above inequality are positive. their sum is positive, and it is proved.

Theorem 5.12. $I_{Kullback}$ satisfies P1.

Proof. We need to show that

$$log_2 m - \sum_{i=1}^{m} \left(\frac{n_i}{N} \right) \log_2 \left(\frac{m n_i}{N} \right) >$$

$$log_2 m - \left(\left(\frac{n_1 + c}{N} \right) \log_2 \left(\frac{m(n_1 + c)}{N} \right) \right.$$

$$\left. + \sum_{i=2}^{m} \left(\frac{n_i - d_i}{N} \right) \log_2 \left(\frac{m(n_i - d_i)}{N} \right) \right).$$

Canceling terms and multiplying both sides of the above inequality by -1 yields

$$\sum_{i-1}^{m} \left(\frac{n_i}{N} \right) \log_2 \left(\frac{m n_i}{N} \right) <$$

$$\left(\frac{n_1 + c}{N} \right) \log_2 \left(\frac{m(n_1 + c)}{N} \right) + \sum_{i=2}^{m} \left(\frac{n_i - d_i}{N} \right) \log_2 \left(\frac{m(n_i - d_i)}{N} \right).$$

The values in the vector $(n_1, \ldots, n_m)$ are uniformly distributed, so $n_1 = \ldots = n_m$. Substituting n_1 for n_i yields

$$\sum_{i=1}^{m} \left(\frac{n_1}{N} \right) \log_2 \left(\frac{m n_1}{N} \right) <$$

$$\left(\frac{n_1 + c}{N} \right) \log_2 \left(\frac{m(n_1 + c)}{N} \right) + \sum_{i=2}^{m} \left(\frac{n_1 - d_i}{N} \right) \log_2 \left(\frac{m(n_1 - d_i)}{N} \right).$$

Factoring $1/N$ out of both sides of the above inequality yields

$$\sum_{i=1}^{m} n_1 \log_2 \left(\frac{m n_1}{N} \right) <$$

$$(n_1 + c) \log_2 \left(\frac{m(n_1 + c)}{N} \right) + \sum_{i=2}^{m} (n_1 - d_i) \log_2 \left(\frac{m(n_1 - d_i)}{N} \right).$$

Using the difference property of logarithms to decompose the above inequality yields

$$\sum_{i=1}^{m} n_1 \log_2(mn_1) - \sum_{i=1}^{m} n_1 \log_2 N <$$

$$(n_1 + c) \log_2(m(n_1 + c)) - n_1 \log_2 N - c \log_2 N$$

$$+ \sum_{i=2}^{m} (n_1 - d_i) \log_2(m(n_1 - d_i)) - \sum_{i=2}^{m} n_1 \log_2 N + \sum_{i=2}^{m} d_i \log_2 N.$$

Canceling terms and using the addition property of logarithms to further decompose the above inequality yields

$$\sum_{i=1}^{m} n_1 \log_2 m + \sum_{i=1}^{m} n_1 \log_2 n_1 <$$

$$n_1 \log_2 m + c \log_2 m + (n_1 + c) \log_2(n_1 + c) + \sum_{i=2}^{m} n_1 \log_2 m$$

$$- \sum_{i=2}^{m} d_i \log_2 m + \sum_{i=2}^{m} (n_1 - d_i) \log_2(n_1 - d_i).$$

Canceling terms yields

$$\sum_{i=1}^{m} n_1 \log_2 n_1 < (n_1 + c) \log_2(n_1 + c) + \sum_{i=2}^{m} (n_1 - d_i) \log_2(n_1 - d_i),$$

which is proved in Theorem 5.3.

Theorem 5.13. $I_{MacArthur}$ satisfies P1.

Proof. We need to show that

$$-\sum_{i=1}^{m} \left(\frac{n_i + \bar{u}}{2N}\right) \log_2 \left(\frac{n_i + \bar{u}}{2N}\right) - \frac{\left(-\sum_{i=1}^{m} \left(\frac{n_i}{N}\right) \log_2 \left(\frac{n_i}{N}\right)\right) + \log_2 m}{2} = 0.$$

The values in the vector $(n_1, \ldots, n_m)$ are uniformly distributed, so $n_i = \bar{u}$. Substituting for $\bar{u}$ in the above equality yields

$$-\sum_{i=1}^{m} \left(\frac{2n_i}{2N}\right) \log_2 \left(\frac{2n_i}{2N}\right) - \frac{\left(-\sum_{i=1}^{m} \left(\frac{n_i}{N}\right) \log_2 \left(\frac{n_i}{N}\right)\right) + \log_2 m}{2} = 0.$$

Also, the values in the vector $(n_1, \ldots, n_m)$ are uniformly distributed, so $n_i/N = 1/m$. Substituting for n_i/N in the above equality yields

$$-\sum_{i=1}^{m} \left(\frac{1}{m}\right) \log_2 \left(\frac{1}{m}\right) - \frac{\left(-\sum_{i=1}^{m} \left(\frac{1}{m}\right) \log_2 \left(\frac{1}{m}\right)\right) + \log_2 m}{2} = 0.$$

Now $\sum_{i=1}^{m} 1/m = 1$, so substituting for $\sum_{i=1}^{m} 1/m$ in the above equality yields

$$-\log_2\left(\frac{1}{m}\right) - \frac{\left(-\log_2\left(\frac{1}{m}\right)\right) + \log_2 m}{2} = 0.$$

Using the difference property of logarithms, we have $\log_2 1/m = \log_2 1 - \log_2 m$, and $\log_2 1 = 0$, so decomposing the above equality and canceling terms yields

$$\log_2 m - \frac{\log_2 m + \log_2 m}{2} = \log_2 m - \frac{2\log_2 m}{2} = \log_2 m - \log_2 m = 0,$$

and it is proved.

Theorem 5.14. I_{Theil} satisfies P1.

Proof. We need to show that

$$\sum_{i=1}^{m} \left| \left(\frac{n_i}{N}\right)\log_2\left(\frac{n_i}{N}\right) - \left(\frac{1}{m}\right)\log_2\left(\frac{1}{m}\right) \right| = 0.$$

The values in the vector $(n_1, \ldots, n_m)$ are uniformly distributed, so $n_i/N = 1/m$. Substituting for n_i/N in the above equality yields

$$\sum_{i=1}^{m} \left| \left(\frac{1}{m}\right)\log_2\left(\frac{1}{m}\right) - \left(\frac{1}{m}\right)\log_2\left(\frac{1}{m}\right) \right| = \sum_{i=1}^{m} |0| = 0,$$

and it is proved.

Theorem 5.15. $I_{Atkinson}$ satisfies P1.

Proof. We need to show that

$$1 - \left(\prod_{i=1}^{m} \frac{\left(\frac{n_i}{N}\right)}{\left(\frac{1}{m}\right)}\right)^{\frac{1}{m}} = 0.$$

The values in the vector $(n_1, \ldots, n_m)$ are uniformly distributed, so $n_i/N = 1/m$. Substituting for n_i/N in the above equality yields

$$1 - \left(\prod_{i=1}^{m} \frac{\left(\frac{1}{m}\right)}{\left(\frac{1}{m}\right)}\right)^{\frac{1}{m}} = 1 - \left(\prod_{i=1}^{m} 1\right)^{\frac{1}{m}} = 1 - (1)^{\frac{1}{m}} = 1 - 1 = 0,$$

and it is proved.

Theorem 5.16. I_{Max} does not satisfy P1.

Proof. Let $N = 50$, $m = 2$, $n = (n_1, n_2) = (25, 25)$, and $n' = (n_1', n_2') = (49, 1)$.

5.3.2. Maximum Value Principle

Maximum Value Principle (P2). Given a vector $(n_1, \ldots, n_m)$, where $n_1 = N - m + 1$, $n_i = 1$, $i = 2, \ldots, m$, and $N > m$, $f(n_1, \ldots, n_m)$ attains its maximum value.

We need to show that $f(n_1, \ldots, n_m) > f(n'_1, \ldots, n'_m)$ for all vectors $(n'_1, \ldots, n'_m)$, where $n'_1 < n_1$. Now for some $c > 0$, we have $n'_1 + c = n_1$, and for each $d_i \geq 0$, where $\sum_{i=2}^m d_i = c$, we have $n'_i - d_i = 1$. So, showing that $f(n_1, \ldots, n_m) > f(n'_1, \ldots, n'_m)$ is equivalent to showing that $f(n'_1 + c, n'_2 - d_2, \ldots, n'_m - d_m) > f(n'_1, \ldots, n'_m)$. For simplicity, we drop the prime character (i.e., $'$) in the proofs that follow and refer to n'_i as n_i.

Theorem 5.17. $I_{Variance}$ satisfies P2.

Proof. We need to show that

$$\left(\frac{n_1 + c}{N} - \frac{1}{m}\right)^2 + \sum_{i=2}^m \left(\frac{n_i - d_i}{N} - \frac{1}{m}\right)^2 > \left(\frac{n_1}{N} - \frac{1}{m}\right)^2 + \sum_{i=2}^m \left(\frac{n_i}{N} - \frac{1}{m}\right)^2,$$

which is equivalent to showing that

$$\left(\frac{n_1 + c}{N} - \frac{1}{m}\right)^2 + \sum_{i=2}^m \left(\frac{n_i - d_i}{N} - \frac{1}{m}\right)^2 - \left(\frac{n_1}{N} - \frac{1}{m}\right)^2$$
$$- \sum_{i=2}^m \left(\frac{n_i}{N} - \frac{1}{m}\right)^2 > 0.$$

Expanding the left side of the above inequality, putting it over the common denominator Nm, and canceling terms yields

$$2n_1 m^2 c - 2mcN + m^2 c^2 + \sum_{i=2}^m (2md_i N - 2n_i m^2 d_i + m^2 d_i^2) > 0.$$

Since $c = \sum_{i=2}^m d_i$, substituting for c in the above inequality yields

$$2n_1 m^2 \sum_{i=2}^m d_i - 2mN \sum_{i=2}^m d_i + m^2 \left(\sum_{i=2}^m d_i\right)^2 + 2mN \sum_{i=2}^m d_i$$
$$- 2m^2 \sum_{i=2}^m n_i d_i + m^2 \sum_{i=2}^m d_i^2 > 0.$$

Canceling terms and factoring m^2 out of the left side of the above inequality yields

$$2n_1 \sum_{i=2}^m d_i + \left(\sum_{i=2}^m d_i\right)^2 - 2 \sum_{i=2}^m n_i d_i + \sum_{i=2}^m d_i^2 > 0.$$

Now one of the positive terms is greater than the negative term, specifically,

$$2n_1 \sum_{i=2}^{m} d_i > 2 \sum_{i=2}^{m} n_i d_i.$$

The two remaining terms are both positive, their sum is positive, and it is proved.

Theorem 5.18. $I_{Simpson}$ satisfies P2.

Proof. We need to show that

$$\left(\frac{n_1 + c}{N}\right)^2 + \sum_{i=2}^{m} \left(\frac{n_i - d_i}{N}\right)^2 > \left(\frac{n_1}{N}\right)^2 + \sum_{i=2}^{m} \left(\frac{n_i}{N}\right)^2,$$

which is equivalent to showing that

$$\left(\frac{n_1 + c}{N}\right)^2 + \sum_{i=2}^{m} \left(\frac{n_i - d_i}{N}\right)^2 - \left(\frac{n_1}{N}\right)^2 - \sum_{i=2}^{m} \left(\frac{n_i}{N}\right)^2 > 0.$$

Expanding the left side of the above inequality, putting it over the common denominator N^2, and canceling terms yields

$$2n_1 c + c^2 + \sum_{i=2}^{m} (n_i^2 - 2n_i d_i + d_i^2) - \sum_{i=2}^{m} n_i^2 > 0.$$

Since $c = \sum_{i=2}^{m} d_i$, substituting for c in the above inequality yields

$$2n_1 \sum_{i=2}^{m} d_i + \left(\sum_{i=2}^{m} d_i\right)^2 + \sum_{i=2}^{m} n_i^2 - 2 \sum_{i=2}^{m} n_i d_i + \sum_{i=2}^{m} d_i^2 - \sum_{i=2}^{m} n_i^2 > 0.$$

Canceling terms yields

$$2n_1 \sum_{i=2}^{m} d_i + \left(\sum_{i=2}^{m} d_i\right)^2 - 2 \sum_{i=2}^{m} n_i d_i + \sum_{i=2}^{m} d_i^2 > 0.$$

Now one of the positive terms is greater than the negative term, specifically,

$$2n_1 \sum_{i=2}^{m} d_i > 2 \sum_{i=2}^{m} n_i d_i.$$

The two remaining terms are both positive, their sum is positive, and it is proved.

Theorem 5.19. $I_{Shannon}$ satisfies P2.

Proof. We need to show that

$$
-\left(\left(\frac{n_1}{N}\right)\log_2\left(\frac{n_1}{N}\right)+\sum_{i=1}^{m}\left(\frac{n_i}{N}\right)\log_2\left(\frac{n_i}{N}\right)\right) >
$$
$$
-\left(\left(\frac{n_1+c}{N}\right)\log_2\left(\frac{n_1+c}{N}\right)+\sum_{i=2}^{m}\left(\frac{n_i-d_i}{N}\right)\log_2\left(\frac{n_i-d_i}{N}\right)\right).
$$

Multiplying both sides by -1 yields

$$
\left(\frac{n_1}{N}\right)\log_2\left(\frac{n_1}{N}\right)+\sum_{i=1}^{m}\left(\frac{n_i}{N}\right)\log_2\left(\frac{n_i}{N}\right) <
$$
$$
\left(\frac{n_1+c}{N}\right)\log_2\left(\frac{n_1+c}{N}\right)+\sum_{i=2}^{m}\left(\frac{n_i-d_i}{N}\right)\log_2\left(\frac{n_i-d_i}{N}\right).
$$

Factoring $1/N$ out of both sides of the above inequality yields

$$
n_1\log_2\left(\frac{n_1}{N}\right)+\sum_{i=1}^{m}n_i\log_2\left(\frac{n_i}{N}\right) <
$$
$$
(n_1+c)\log_2\left(\frac{n_1+c}{N}\right)+\sum_{i=2}^{m}(n_i-d_i)\log_2\left(\frac{n_i-d_i}{N}\right),
$$

which is equivalent to

$$
n_1\log_2\left(\frac{n_1}{N}\right)+\sum_{i=1}^{m}n_i\log_2\left(\frac{n_i}{N}\right) <
$$
$$
n_1\log_2\left(\frac{n_1+c}{N}\right)+c\log_2\left(\frac{n_1+c}{N}\right)+\sum_{i=2}^{m}n_i\log_2\left(\frac{n_i-d_i}{N}\right)
$$
$$
-d_i\log_2\left(\frac{n_i-d_i}{N}\right).
$$

Using the difference property of logarithms to decompose the above inequality
yields

$$n_1 \log_2 n_1 - n_1 \log_2 N + \sum_{i=1}^{m} n_i \log_2 n_i - \sum_{i=1}^{m} n_i \log_2 N <$$

$$n_1 \log_2(n_1 + c) + c \log_2(n_1 + c) - n_1 \log_2 N - c \log_2 N$$

$$+ \sum_{i=2}^{m} n_i \log_2(n_i - d_i) - \sum_{i=2}^{m} d_i \log_2(n_i - d_i) - \sum_{i=2}^{m} n_i \log_2 N$$

$$+ \sum_{i=2}^{m} d_i \log_2 N.$$

Now $c = \sum_{i=2}^{m} d_i$, so substituting for c on the right side of the above inequality and canceling terms yields

$$n_1 \log_2 n_1 + \sum_{i=1}^{m} n_i \log_2 n_i < (n_1 + c) \log_2(n_1 + c) + \sum_{i=2}^{m} (n_i - d_i) \log_2(n_i - d_i),$$

which can be written as

$$n_1 \log_2 n_1 + n_2 \log_2 n_2 + \ldots + n_m \log_2 n_m <$$
$$(n_1 + c) \log_2(n_1 + c) + (n_2 - d_2) \log_2(n_2 - d_2) + \ldots$$
$$+ (n_m - d_m) \log_2(n_m - d_m).$$

Since the function $x \log_2 x$ is a convex function, we have

$$(n_1 + c) \log_2(n_1 + c) + (n_2 - d_2) \log_2(n_2 - d_2)$$
$$+ \ldots + (n_m - d_m) \log_2(n_m - d_m) \geq$$
$$(n_1 + c + n_2 - d_2 + \ldots + n_m - d_m) \log_2(n_1 + c + n_2 - d_2 + \ldots$$
$$+ n_m - d_m).$$

Now $c = \sum_{i=2}^{m} d_i$, so substituting for the d_i's on the right side of the above inequality yields

$$(n_1 + c + n_1 - d_2 + \ldots + n_1 - d_m) \log_2(n_1 + c + n_1 - d_2 + \ldots$$
$$+ n_1 - d_m) = (n_1 + n_2 + \ldots + n_m) \log_2(n_1 + n_2 + \ldots + n_m).$$

We now show that

$$n_1 \log_2 n_1 + n_2 \log_2 n_2 + \ldots + n_m \log_2 n_m <$$
$$(n_1 + n_2 + \ldots + n_m) \log_2(n_1 + n_2 + \ldots + n_m).$$

Distributing $(n_1 + n_2 + \ldots + n_m)$ over the log on the right side of the above inequality yields

$$n_1 \log_2 n_1 + n_2 \log_2 n_2 + \ldots + n_m \log_2 n_m <$$
$$n_1 \log_2 (n_1 + n_2 + \ldots + n_m) + n_2 \log_2 (n_1 + n_2 + \ldots + n_m) + \ldots$$
$$+ n_m \log_2 (n_1 + n_2 + \ldots + n_m),$$

which is obviously true, and it is proved.

Theorem 5.20. I_{Total} satisfies P2.

Proof. We need to show that

$$m \left(- \sum_{i=1}^{m} \left(\frac{n_i}{N} \right) \log_2 \left(\frac{n_i}{N} \right) \right) >$$
$$m \left(- \left(\left(\frac{n_1 + c}{N} \right) \log_2 \left(\frac{n_1 + c}{N} \right) \right. \right.$$
$$\left. \left. + \sum_{i=2}^{m} \left(\frac{n_i - d_i}{N} \right) \log_2 \left(\frac{n_i - d_i}{N} \right) \right) \right).$$

Canceling terms and multiplying both sides of the above inequality by -1 yields

$$\sum_{i=1}^{m} \left(\frac{n_i}{N} \right) \log_2 \left(\frac{n_i}{N} \right) <$$
$$\left(\frac{n_1 + c}{N} \right) \log_2 \left(\frac{n_1}{N} \right) + \sum_{i=2}^{m} \left(\frac{n_i - d_i}{N} \right) \log_2 \left(\frac{n_i - d_i}{N} \right),$$

which is proved in Theorem 5.19.

Theorem 5.21. $I_{McIntosh}$ satisfies P2.

Proof. We need to show that

$$\frac{N - \sqrt{(n_1 + c)^2 + \sum_{i=2}^{m} (n_i - d_i)^2}}{N - \sqrt{N}} < \frac{N - \sqrt{n_1^2 + \sum_{i=2}^{m} n_i^2}}{N - \sqrt{N}}.$$

Canceling terms and multiplying both sides of the above inequality by -1 yields

$$(n_1 + c)^2 + \sum_{i=2}^{m} (n_i - d_i)^2 > n_1^2 + \sum_{i=2}^{m} n_i^2.$$

Expanding the terms on the left side of the above inequality yields

$$n_1^2 + 2n_1 c + c^2 + \sum_{i=2}^{m} (n_i^2 - 2n_i d_i + d_i^2) > n_1^2 + \sum_{i=2}^{m} n_i^2.$$

Distributing the summation on the left side of the above inequality yields

$$n_1^2 + 2n_1 c + c^2 + \sum_{i=2}^{m} n_i^2 - \sum_{i=2}^{m} 2n_i d_i + \sum_{i=2}^{m} d_i^2 > n_1^2 + \sum_{i=2}^{m} n_i^2.$$

Since $c = \sum_{i=2}^{m} d_i$, substituting for c in the above inequality yields

$$n_1^2 + \sum_{i=2}^{m} 2n_1 d_i + \left(\sum_{i=2}^{m} d_i \right)^2 + \sum_{i=2}^{m} n_i^2 - \sum_{i=2}^{m} 2n_i d_i + \sum_{i=2}^{m} d_i^2 > n_1^2 + \sum_{i=2}^{m} n_i^2.$$

Canceling terms yields

$$2 \sum_{i=2}^{m} (n_1 - n_i) d_i + \left(\sum_{i=2}^{m} d_i \right)^2 + \sum_{i=2}^{m} d_i^2 > 0.$$

Since all terms on the left side of the above inequality are positive, their sum is positive, and it is proved.

Theorem 5.22. I_{Lorenz} satisfies P2.

Proof. We need to show that

$$\frac{1}{m} \left(\sum_{i=1}^{m-1} (m - i + 1) \frac{n_i}{N} + \frac{n_m}{N} \right) >$$

$$\frac{1}{m} \left(\sum_{i=1}^{m-1} (m - i + 1) \frac{n_i - d_i}{N} + \frac{n_m + c}{N} \right).$$

Canceling terms and distributing the summation on the right side of the above inequality yields

$$\sum_{i=1}^{m-1} (m - i + 1) \frac{n_i}{N} + \frac{n_m}{N} > \sum_{i=1}^{m-1} (m - i + 1) \frac{n_i}{N} - \sum_{i=1}^{m-1} (m - i + 1) \frac{d_i}{N} + \frac{n_m}{N} + \frac{c}{N}.$$

Canceling terms yields

$$0 > - \sum_{i=1}^{m-1} (m - i + 1) \frac{d_i}{N} + \frac{c}{N}.$$

Since $c = \sum_{i=1}^{m-1} d_i$, substituting for c in the above inequality, and moving $- \sum_{i=1}^{m-1} \frac{d_i}{N}$ to the left side yields

$$\sum_{i=1}^{m-1} (m - i + 1) \frac{d_i}{N} > \sum_{i=1}^{m-1} \frac{d_i}{N}.$$

Factoring $1/N$ out of both sides of the above inequality yields

$$\sum_{i=1}^{m-1} (m - i + 1)d_i > \sum_{i=1}^{m-1} d_i,$$

which is obviously true, and it is proved.

Theorem 5.23. I_{Gini} satisfies P2.

Proof. To simplify the proof of this theorem, we use an equivalent form of the Gini index described in [31], where it is given as

$$\frac{2\left(\frac{m+1}{2} - \sum_{i=1}^{m} i\left(\frac{n_i}{N}\right)\right)}{m}.$$

We need to show that

$$\frac{2\left(\frac{m+1}{2} - \left(\frac{n_1+c}{N} + \sum_{i=2}^{m} i\left(\frac{n_i-d_i}{N}\right)\right)\right)}{m} > \frac{2\left(\frac{m+1}{2} - \left(\frac{n_1}{N} + \sum_{i=2}^{m} i\left(\frac{n_i}{N}\right)\right)\right)}{m}.$$

Canceling terms and multiplying both sides of the above inequality by -1 yields

$$n_1 + c + \sum_{i=2}^{m} i(n_i - d_i) < n_1 + \sum_{i=2}^{m} i n_i.$$

Distributing the summation on the left side of the above inequality, canceling terms and moving $-\sum_{i=2}^{m} id_i$ to the right side yields

$$c < \sum_{i=2}^{m} id_i.$$

Since $c = \sum_{i=2}^{m} d_i$, substituting for c in the above inequality yields

$$\sum_{i=2}^{m} d_i < \sum_{i=2}^{m} id_i,$$

which is obviously true, and it is proved.

Theorem 5.24. I_{Berger} satisfies P2.

Proof. We need to show that

$$\frac{n_1 + c}{N} > \frac{n_1}{N},$$

which is proved in Theorem 5.8.

Theorem 5.25. I_{Schutz} satisfies P2.

Proof. We need to show that

$$\left| \frac{N-m+1}{N} - \frac{1}{m} \right| + \sum_{i=2}^{m} \left| \frac{1}{N} - \frac{1}{m} \right| > \sum_{i=1}^{m} \left| \frac{n_i}{N} - \frac{1}{m} \right|,$$

which is equivalent to showing that

$$\left| \frac{N-m+1}{N} - \frac{1}{m} \right| + \sum_{i=2}^{m} \left| \frac{1}{N} - \frac{1}{m} \right| > \sum_{i=1}^{k} \left| \frac{n_i}{N} - \frac{1}{m} \right| + \sum_{i=k+1}^{m} \left| \frac{n_i}{N} - \frac{1}{m} \right|.$$

If we assume that k is chosen such that $n_i/N > 1/m$, for $i \leq k$, and $n_i/N < 1/m$, for $i > k$, then the above inequality is equivalent to

$$\left(\frac{N-m+1}{N} - \frac{1}{m} \right) + \sum_{i=2}^{m} \left(\frac{1}{N} - \frac{1}{m} \right) >$$

$$\sum_{i=1}^{k} \left(\frac{n_i}{N} - \frac{1}{m} \right) - \sum_{i=k+1}^{m} \left(\frac{n_i}{N} - \frac{1}{m} \right).$$

Now for every $k \in \{1, \ldots, m\}$,

$$\sum_{i=1}^{k} \left(\frac{n_i}{N} - \frac{1}{m} \right) = - \sum_{i=k+1}^{m} \left(\frac{n_i}{N} - \frac{1}{m} \right),$$

so substituting for $- \sum_{i=k+1}^{m} (n_i/N - 1/m)$ in the above inequality yields

$$2 \left(\frac{N-m+1}{N} - \frac{1}{m} \right) > 2 \sum_{i=1}^{k} \left(\frac{n_i}{N} - \frac{1}{m} \right).$$

Canceling terms and putting both sides of the above inequality over the common denominator Nm yields

$$Nm - m^2 + m - N > \sum_{i=1}^{k} mn_i - Nk,$$

which is equivalent to

$$Nm + Nk - N > \sum_{i=1}^{k} mn_i + m^2 - m.$$

Since $N = mn_k$ (i.e., n_k is the average), substituting for N in the above inequality and factoring m out of both sides yields

$$mn_k + kn_k - n_k > \sum_{i=1}^{k} n_i + m - 1,$$

which is equivalent to

$$mn_k + kn_k + 1 > \sum_{i=1}^{k} n_i + m + n_k.$$

Since $mn_k = N = \sum_{i=1}^{m} n_i = \sum_{i=1}^{k} n_i + \sum_{i=k+1}^{m} n_i$, substituting for mn_k in the above inequality yields

$$\sum_{i=1}^{k} n_i + \sum_{i=k+1}^{m} n_i + kn_k + 1 > \sum_{i=1}^{k} n_i + m + n_k.$$

Canceling $\sum_{i=1}^{k} n_i$ from both sides of the above inequality, and moving n_k to the left side, yields

$$\sum_{i=k+1}^{m} n_i + (k-1)n_k + 1 > m,$$

which can be written as

$$n_k + n_k + \ldots + n_k + n_{k+1} + n_{k+2} + \ldots + n_m + 1 > m.$$

The left side of the above inequality contains m terms, consisting of $(k-1)$ n_k terms, $(m-k)$ n_i terms, and 1. The vector $(n_1, \ldots, n_m)$ is not a uniform distribution, so $n_k > 1$, each $n_i \geq 1$, for $i = k+1, \ldots, m$, the sum of these terms is greater than m, and it is proved.

Theorem 5.26. I_{Bray} satisfies P2.

Proof. We need to show that

$$\frac{1}{N}\left(\frac{N}{m} + \sum_{i=2}^{m} 1\right) < \frac{1}{N}\left(\frac{N}{m} + \sum_{i=2}^{k} \frac{N}{m} + \sum_{i=k+1}^{m} (n_i + d_i)\right).$$

Canceling terms yields

$$\sum_{i=2}^{m} 1 < \sum_{i=2}^{k} \frac{N}{m} + \sum_{i=k+1}^{m} (n_i + d_i),$$

which can be written as

$$\sum_{i=2}^{k} 1 + \sum_{i=k+1}^{m} 1 < \sum_{i=2}^{k} \frac{N}{m} + \sum_{i=k+1}^{m} (n_i + d_i).$$

Now $N/m > 1$,

$$\sum_{i=2}^{k} \frac{N}{m} > \sum_{i=2}^{k} 1$$

$n_i + d_i > 1$,

$$\sum_{i=k+1}^{m} (n_i + d_i) > \sum_{i=k+1}^{m} 1,$$

and it is proved.

Theorem 5.27. $I_{Whittaker}$ satisfies P2.

Proof. We need to show that

$$1 - \left(0.5 \sum_{i=1}^{m} \left| \frac{n_i}{N} - \frac{1}{m} \right| \right) >$$

$$1 - \left(0.5 \left(\left| \frac{N - m + 1}{N} - \frac{1}{m} \right| + \sum_{i=2}^{m} \left| \frac{1}{N} - \frac{1}{m} \right| \right) \right).$$

Canceling terms and multiplying both sides of the above inequality by - 1 yields

$$\sum_{i=1}^{m} \left| \frac{n_i}{N} - \frac{1}{m} \right| < \left| \frac{N - m + 1}{N} - \frac{1}{m} \right| + \sum_{i=2}^{m} \left| \frac{1}{N} - \frac{1}{m} \right|,$$

which is proved in Theorem 5.25.

Theorem 5.28. $I_{Kullback}$ satisfies P2.

Proof. We need to show that

$$log_2 m - \left(\left(\frac{n_1}{N} \right) \log_2 \left(\frac{m n_1}{N} \right) + \sum_{i=2}^{m} \left(\frac{n_i}{N} \right) \log_2 \left(\frac{m n_i}{N} \right) \right) >$$

$$log_2 m - \left(\left(\frac{n_1 + c}{N} \right) \log_2 \left(\frac{m(n_1 + c)}{N} \right) \right.$$

$$\left. + \sum_{i=2}^{m} \left(\frac{n_i - d_i}{N} \right) \log_2 \left(\frac{m(n_i - d_i)}{N} \right) \right).$$

Canceling terms and multiplying both sides of the above inequality by -1 yields

$$\left(\frac{n_1}{N} \right) \log_2 \left(\frac{m n_1}{N} \right) + \sum_{i=2}^{m} \left(\frac{n_i}{N} \right) \log_2 \left(\frac{m n_i}{N} \right) <$$

$$\left(\frac{n_1 + c}{N} \right) \log_2 \left(\frac{m(n_1 + c)}{N} \right) + \sum_{i=2}^{m} \left(\frac{n_i - d_i}{N} \right) \log_2 \left(\frac{m(n_i - d_i)}{N} \right).$$

Factoring $1/N$ out of both sides of the above inequality yields

$$n_1 \log_2 \left(\frac{mn_1}{N}\right) + \sum_{i=2}^{m} n_1 \log_2 \left(\frac{mn_1}{N}\right) <$$
$$(n_1 + c) \log_2 \left(\frac{m(n_1 + c)}{N}\right) + \sum_{i=2}^{m} (n_i - d_i) \log_2 \left(\frac{m(n_i - d_i)}{N}\right),$$

which is equivalent to

$$n_1 \log_2 \left(\frac{mn_1}{N}\right) + \sum_{i=2}^{m} n_1 \log_2 \left(\frac{mn_1}{N}\right) <$$
$$n_1 \log_2 \left(\frac{m(n_1 + c)}{N}\right) + c \log_2 \left(\frac{m(n_1 + c)}{N}\right)$$
$$+ \sum_{i=2}^{m} n_i \log_2 \left(\frac{m(n_i - d_i)}{N}\right) - \sum_{i=2}^{m} d_i \log_2 \left(\frac{m(n_i - d_i)}{N}\right).$$

Using the difference property of logarithms to decompose the above inequality yields

$$n_1 \log_2 mn_1 - n_1 \log_2 N + \sum_{i=2}^{m} n_i \log_2(mn_i) - \sum_{i=2}^{m} n_i \log_2 N <$$
$$(n_1 + c) \log_2(m(n_1 + c)) - n_1 \log_2 N - c \log_2 N$$
$$+ \sum_{i=2}^{m} n_i \log_2(m(n_i - d_i)) - \sum_{i=2}^{m} d_i \log_2(m(n_i - d_i))$$
$$- \sum_{i=2}^{m} n_i \log_2 N + \sum_{i=2}^{m} d_i \log_2 N.$$

Now $c = \sum_{i=2}^{m} d_i$, so substituting for c on the right side of the above inequality and canceling terms yields

$$n_1 \log_2 mn_1 + \sum_{i=2}^{m} n_i \log_2(mn_i) <$$
$$n_1 \log_2(m(n_1 + c)) + c \log_2(m(n_1 + c)) + \sum_{i=2}^{m} n_i \log_2(m(n_i - d_i))$$
$$- \sum_{i=2}^{m} d_i \log_2(m(n_i - d_i)).$$

Using the addition property of logarithms to further decompose the above inequality yields

$$n_1 \log_2 m + n_1 \log_2 n_1 + \sum_{i=2}^{m} n_i \log_2 m + \sum_{i=2}^{m} n_i \log_2 n_i <$$

$$n_1 \log_2 m + n_1 \log_2 (n_1 + c) + c \log_2 m + c \log_2 (n_1 + c)$$

$$+ \sum_{i=2}^{m} n_i \log_2 m + \sum_{i=2}^{m} n_i \log_2 (n_i - d_i) - \sum_{i=2}^{m} d_i \log_2 m$$

$$- \sum_{i=2}^{m} d_i \log_2 (n_i - d_i).$$

Again substituting for c on the right side of the above inequality, and canceling terms yields

$$n_1 \log_2 n_1 + \sum_{i=2}^{m} n_i \log_2 n_i < (n_1 + c) \log_2 (n_1 + c) + \sum_{i=2}^{m} (n_i - d_i) \log_2 (n_i - d_i),$$

which is proved in Theorem 5.3.

Theorem 5.29. $I_{MacArthur}$ satisfies P2.

Proof. Let the vector $(a, \ldots, a)$ containing m a's be the uniform distribution. Then the vector obtained by combining the vectors $(n_1 + c, n_2 - d_2, \ldots, n_m - d_m)$ and $(a, \ldots, a)$ is the vector $(n_1 + c + a, n_2 - d_2 + a, \ldots, n_m - d_m + a)$, and the vector obtained by combining the vectors $(n_1, \ldots, n_m)$ and $(a, \ldots, a)$ is the vector $(n_1 + a, \ldots, n_m + a)$. We need to show that

$$-\left(\left(\frac{n_1 + c + a}{2N} \right) \log_2 \left(\frac{n_1 + c + a}{2N} \right) \right.$$

$$+ \sum_{i=2}^{m} \left(\frac{n_i - d_i + a}{2N} \right) \log_2 \left. \left(\frac{n_i - d_i + a}{2N} \right) \right)$$

$$- \left(\frac{ -\left(\left(\frac{n_1 + c}{N} \right) \log_2 \left(\frac{n_1 + c}{N} \right) + \sum_{i=2}^{m} \left(\frac{n_i - d_i}{N} \right) \log_2 \left(\frac{n_i - d_i}{N} \right) \right) + \log_2 m }{2} \right) >$$

$$- \left(\left(\frac{n_1 + a}{2N} \right) \log_2 \left(\frac{n_1 + a}{2N} \right) + \sum_{i=2}^{m} \left(\frac{n_i + a}{2N} \right) \log_2 \left(\frac{n_i + a}{2N} \right) \right)$$

$$- \left(\frac{ -\left(\left(\frac{n_1}{N} \right) \log_2 \left(\frac{n_1}{N} \right) + \sum_{i=2}^{m} \left(\frac{n_i}{N} \right) \log_2 \left(\frac{n_i}{N} \right) \right) + \log_2 m }{2} \right),$$

which is equivalent to showing that

$$- \left(\frac{n_1 + c + a}{2N} \right) \log_2 \left(\frac{n_1 + c + a}{2N} \right)$$

$$- \sum_{i=2}^{m} \left(\frac{n_i - d_i + a}{2N} \right) \log_2 \left(\frac{n_i - d_i + a}{2N} \right)$$

$$+ \left(\frac{n_1 + c}{2N} \right) \log_2 \left(\frac{n_1 + c}{N} \right) + \sum_{i=2}^{m} \left(\frac{n_i - d_i}{2N} \right) \log_2 \left(\frac{n_i - d_i}{N} \right)$$

$$- \frac{\log_2 m}{2} >$$

$$- \left(\frac{n_1 + a}{2N} \right) \log_2 \left(\frac{n_1 + a}{2N} \right) - \sum_{i=2}^{m} \left(\frac{n_i + a}{2N} \right) \log_2 \left(\frac{n_i + a}{2N} \right)$$

$$+ \left(\frac{n_1}{2N} \right) \log_2 \left(\frac{n_1}{N} \right) + \sum_{i=2}^{m} \left(\frac{n_i}{2N} \right) \log_2 \left(\frac{n_i}{N} \right) - \frac{\log_2 m}{2}.$$

Canceling terms, factoring $1/2N$ out of both sides of the above inequality, and moving those terms preceded by a minus sign to the other side of the inequality yields

$$(n_1 + a) \log_2 \left(\frac{n_1 + a}{2N} \right) + \sum_{i=2}^{m} (n_i + a) \log_2 \left(\frac{n_i + a}{2N} \right)$$

$$+ (n_1 + c) \log_2 \left(\frac{n_1 + c}{N} \right) + \sum_{i=2}^{m} (n_i - d_i) \log_2 \left(\frac{n_i - d_i}{N} \right) >$$

$$(n_1 + c + a) \log_2 \left(\frac{n_1 + c + a}{2N} \right)$$

$$+ \sum_{i=2}^{m} (n_i - d_i + a) \log_2 \left(\frac{n_i - d_i + a}{2N} \right)$$

$$+ n_1 \log_2 \left(\frac{n_1}{N} \right) + \sum_{i=2}^{m} n_i \log_2 \left(\frac{n_i}{N} \right).$$

Using the difference property of logarithms to decompose the above inequality yields

$$(n_1 + a) \log_2(n_1 + a) - n_1 \log_2(2N) - a \log_2(2N)$$

$$+ \sum_{i=2}^{m} (n_i + a) \log_2(n_i + a) - \sum_{i=2}^{m} n_i \log_2(2N) - \sum_{i=2}^{m} a \log_2(2N)$$

$$+ (n_1 + c) \log_2(n_1 + c) - n_1 \log_2 N - c \log_2 N$$

$$+ \sum_{i=2}^{m} (n_i - d_i) \log_2(n_i - d_i) - \sum_{i=2}^{m} n_i \log_2 N + \sum_{i=2}^{m} d_i \log_2 N >$$

$$(n_1 + c + a) \log_2(n_1 + c + a) - n_1 \log_2(2N) - c \log_2(2N)$$

$$- a \log_2(2N) + \sum_{i=2}^{m} (n_i - d_i + a) \log_2(n_i - d_i + a) - \sum_{i=2}^{m} n_i \log_2(2N)$$

$$+ \sum_{i=2}^{m} d_i \log_2(2N) - \sum_{i=2}^{m} a \log_2(2N) + n_1 \log_2 n_1 - n_1 \log_2 N$$

$$+ \sum_{i=2}^{m} n_i \log_2 n_i - \sum_{i=2}^{m} n_i \log_2 N.$$

Canceling terms yields

$$(n_1 + a) \log_2(n_1 + a) + \sum_{i=2}^{m} (n_i + a) \log_2 \left(\frac{n_i + a}{2N} \right)$$

$$+ (n_1 + c) \log_2 \left(\frac{n_1 + c}{N} \right) + \sum_{i=2}^{m} (n_i - d_i) \log_2 \left(\frac{n_i - d_i}{N} \right) >$$

$$(n_1 + c + a) \log_2 \left(\frac{n_1 + c + a}{2N} \right)$$

$$+ \sum_{i=2}^{m} (n_i - d_i + a) \log_2 \left(\frac{n_i - d_i + a}{2N} \right)$$

$$+ \sum_{i=2}^{m} n_i \log_2 n_i.$$

To show that the above inequality is true, we need to show that

$$(n_1 + a) \log_2(n_1 + a) + (n_1 + c) \log_2(n_1 + c) > (n_1 + c + a) \log_2(n_1 + c + a)$$

and

$$\sum_{i=2}^{m} (n_i + a) \log_2(n_i + a) + \sum_{i=2}^{m} (n_i - d_i) \log_2(n_i - d_i) >$$

$$\sum_{i=2}^{m} (n_i - d_i + a) \log_2(n_i - d_i + a) + \sum_{i=2}^{m} n_i \log_2 n_i.$$

We now show that

$$(n_1 + a) \log_2(n_1 + a) + (n_1 + c) \log_2(n_1 + c) >$$
$$(n_1 + c + a) \log_2(n_1 + c + a).$$

Since the function $x \log_2 x$ is a convex function, we have

$$(n_1 + a) \log_2(n_1 + a) + (n_1 + c) \log_2(n_1 + c) \geq$$
$$(n_1 + c + n_1 + a) \log_2(n_1 + c + n_1 + a),$$

so obviously

$$(n_1 + a) \log_2(n_1 + a) + (n_1 + c) \log_2(n_1 + c) >$$
$$(n_1 + c + a) \log_2(n_1 + c + a),$$

and the first part is shown. We now show that

$$\sum_{i=2}^{m}(n_i + a) \log_2(n_i + a) + \sum_{i=2}^{m}(n_i - d_i) \log_2(n_i - d_i) >$$
$$\sum_{i=2}^{m}(n_i - d_i + a) \log_2(n_i - d_i + a) + \sum_{i=2}^{m} n_i \log_2 n_i.$$

The above inequality can be written as

$$(n_2 + a) \log_2(n_2 + a) + (n_3 + a) \log_2(n_3 + a) + \ldots$$
$$+(n_m + a) \log_2(n_m + a) + (n_2 - d_2) \log_2(n_2 - d_2)$$
$$+(n_3 - d_3) \log_2(n_3 - d_3) + \ldots + (n_m - d_m) \log_2(n_m - d_m) >$$
$$(n_2 - d_2 + a) \log_2(n_2 - d_2 + a)$$
$$+(n_3 - d_3 + a) \log_2(n_3 - d_3 + a) + \ldots$$
$$+(n_m - d_m + a) \log_2(n_m - d_m + a) + n_2 \log_2 n_2$$
$$+n_3 \log_2 n_3 + \ldots + n_m \log_2 n_m.$$

Since the function $x \log_2 x$ is a convex function, we have

$$(n_2 + a) \log_2(n_2 + a) + (n_3 + a) \log_2(n_3 + a) + \ldots$$
$$+(n_m + a) \log_2(n_m + a) + (n_2 - d_2) \log_2(n_2 - d_2)$$
$$+(n_3 - d_3) \log_2(n_3 - d_3) + \ldots + (n_m - d_m) \log_2(n_m - d_m) \geq$$
$$(n_2 + a + n_3 + a + \ldots + n_m + a + n_2 - d_2 + n_3 - d_3 + \ldots$$
$$+n_m - d_m)$$
$$\log_2(n_2 + a + n_3 + a + \ldots + n_m + a + n_2 - d_2 + n_3 - d_3 + \ldots$$
$$+n_m - d_m),$$

which can be written as

$$(n_2 + a) \log_2(n_2 + a) + (n_3 + a) \log_2(n_3 + a) + \ldots$$
$$+ (n_m + a) \log_2(n_m + a) + (n_2 - d_2) \log_2(n_2 - d_2)$$
$$+ (n_3 - d_3) \log_2(n_3 - d_3) + \ldots + (n_m - d_m) \log_2(n_m - d_m) \geq$$
$$(n_2 + n_3 + \ldots + n_m + (n_2 - d_2 + a) + (n_3 - d_3 + a) + \ldots$$
$$+ (n_m - d_m + a)) \log_2(n_2 + n_3 + \ldots + n_m + (n_2 - d_2 + a)$$
$$+ (n_3 - d_3 + a) + \ldots + (n_m - d_m + a)).$$

Distributing $(n_2 + n_3 + \ldots + n_m + (n_2 - d_2 + a) + (n_3 - d_3 + a) + \ldots + (n_m - d_m + a))$ over the log in the above inequality yields

$$(n_2 + a) \log_2(n_2 + a) + (n_3 + a) \log_2(n_3 + a) + \ldots$$
$$+ (n_m + a) \log_2(n_m + a) + (n_2 - d_2) \log_2(n_2 - d_2)$$
$$+ (n_3 - d_3) \log_2(n_3 - d_3) + \ldots + (n_m - d_m) \log_2(n_m - d_m) \geq$$
$$(n_2 \log_2(n_2 + n_3 + \ldots + n_m + (n_2 - d_2 + a) + (n_3 - d_3 + a) + \ldots$$
$$+ (n_m - d_m + a)) + n_3 \log_2(n_2 + n_3 + \ldots + n_m + (n_2 - d_2 + a)$$
$$+ (n_3 - d_3 + a) + \ldots + (n_m - d_m + a)) + \ldots$$
$$+ n_m \log_2(n_2 + n_3 + \ldots + n_m + (n_2 - d_2 + a) + (n_3 - d_3 + a) + \ldots$$
$$+ (n_m - d_m + a)) + (n_2 - d_2 + a) \log_2(n_2 + n_3 + \ldots + n_m$$
$$+ (n_2 - d_2 + a) + (n_3 - d_3 + a) + \ldots + (n_m - d_m + a))$$
$$+ (n_3 - d_3 + a) \log_2(n_2 + n_3 + \ldots + n_m + (n_2 - d_2 + a)$$
$$+ (n_3 - d_3 + a) + \ldots + (n_m - d_m + a)) + \ldots$$
$$+ (n_m - d_m + a)) \log_2(n_2 + n_3 + \ldots + n_m + (n_2 - d_2 + a)$$
$$+ (n_3 - d_3 + a) + \ldots + (n_m - d_m + a))$$

so obviously

$$(n_2 + a) \log_2(n_2 + a) + (n_3 + a) \log_2(n_3 + a) + \ldots$$
$$+ (n_m + a) \log_2(n_m + a) + (n_2 - d_2) \log_2(n_2 - d_2)$$
$$+ (n_3 - d_3) \log_2(n_3 - d_3) + \ldots + (n_m - d_m) \log_2(n_m - d_m) >$$
$$n_2 \log_2 n_2 + n_3 \log_2 n_3 + \ldots + n_m \log_2 n_m$$
$$+ (n_2 - d_2 + a) \log_2(n_2 - d_2 + a)$$
$$+ (n_3 - d_3 + a) \log_2(n_3 - d_3 + a) + \ldots$$
$$+ (n_m - d_m + a) \log_2(n_m - d_m + a),$$

the second part is shown, and it is proved.

Theorem 5.30. $I_{Atkinson}$ satisfies P2.

Proof. We need to show that

$$1 - \left(\frac{\left(\frac{N-m+1}{N} \right)}{\left(\frac{1}{m} \right)} \prod_{i=2}^{m} \frac{\left(\frac{1}{N} \right)}{\left(\frac{1}{m} \right)} \right)^{\frac{1}{m}} > 1 - \left(\prod_{i=1}^{m} \frac{\left(\frac{n_i}{N} \right)}{\left(\frac{1}{m} \right)} \right)^{\frac{1}{m}}.$$

Canceling terms and multiplying both sides by -1 yields

$$N - m + 1 < \prod_{i=1}^{m} n_i,$$

which is equivalent to

$$N + 1 < \prod_{i=1}^{m} n_i + m.$$

Since $N = \sum_{i=1}^{m} n_i$, substituting for N in the above inequality yields

$$\sum_{i=1}^{m} n_i + 1 < \prod_{i=1}^{m} n_i + m.$$

The vector $(n_1, \ldots, n_m)$ is not a uniform distribution, so $n_1 > 1$ and $\sum_{i=1}^{m} n_i < \prod_{i=1}^{m} n_i$. Also, $1 < m$, so the sum of the terms on the left side of the above inequality is less than the sum of the terms on the right side, and it is proved.

Theorem 5.31. I_{Max} and I_{Theil} do not satisfy P2.

Proof. Let $N = 50$, $m = 2$, $n = (n_1, n_2) = (49, 1)$, and $n' = (n'_1, n'_2) = (25, 25)$.

5.3.3. Skewness Principle

Skewness Principle (P3). Given a vector $(n_1, \ldots, n_m)$, where $n_1 = N - m + 1$, $n_i = 1, i = 2, \ldots, m$, and $N > m$, and a vector $(n_1 - c, n_2, \ldots, n_m, n_{m+1}, \ldots, n_{m+c})$, where $n_1 - c > 1$ and $n_i = 1, i = 2, \ldots, m + c$, $f(n_1, \ldots, n_m) > f(n_1 - c, n_2, \ldots, n_m, n_{m+1}, \ldots, n_{m+c})$.

Conjecture 5.32. $I_{Variance}$ satisfies P3.

Proof. This has been shown to be true in all experiments conducted using $I_{Variance}$. The proof involves polynomials of many terms (i.e., more than 80) that are not easily simplified, even with the assistance of an automated symbolic computation system. Consequently, the proof will be completed as part of future research.

Theorem 5.33. $I_{Simpson}$ satisfies P3.

Proof. We need to show that

$$\left(\frac{n_1}{N}\right)^2 + \sum_{i=2}^{m}\left(\frac{1}{N}\right)^2 > \left(\frac{n_1-c}{N}\right)^2 + \sum_{i=2}^{m+c}\left(\frac{1}{N}\right)^2.$$

Expanding both sides of the above inequality yields

$$\frac{n_1^2}{N^2} + \frac{m-1}{N^2} > \frac{n_1^2 - 2n_1c + c^2}{N^2} + \frac{m+c-1}{N^2}.$$

Canceling terms yields

$$\frac{2n_1c}{N^2} > \frac{c^2+c}{N^2}.$$

Factoring c/N^2 out of both sides of the above inequality yields

$$2n_1 > c+1.$$

Now $c \geq 1$ and $n_1 > c$, so $2n_1 > c+1$, and it is proved.

Theorem 5.34. $I_{Shannon}$ satisfies P3.

Proof. We need to show that

$$-\left(\left(\frac{n_1}{N}\right)\log_2\left(\frac{n_1}{N}\right) + \sum_{i=2}^{m}\left(\frac{1}{N}\right)\log_2\left(\frac{1}{N}\right)\right) <$$
$$-\left(\left(\frac{n_1-c}{N}\right)\log_2\left(\frac{n_1-c}{N}\right) + \sum_{i=2}^{m+c}\left(\frac{1}{N}\right)\log_2\left(\frac{1}{N}\right)\right).$$

Multiplying both sides of the above inequality by -1 and factoring $1/N$ out of both sides yields

$$n_1\log_2\left(\frac{n_1}{N}\right) + (m-1)\log_2\left(\frac{1}{N}\right) >$$
$$(n_1-c)\log_2\left(\frac{n_1-c}{N}\right) + (m+c-1)\log_2\left(\frac{1}{N}\right),$$

which is equivalent to

$$n_1\log_2\left(\frac{n_1}{N}\right) > (n_1-c)\log_2\left(\frac{n_1-c}{N}\right) + c\log_2\left(\frac{1}{N}\right).$$

Using the difference and addition properties of logarithms yields

$$n_1 \log_2 n_1 - n_1 \log_2 N >$$
$$n_1 \log_2(n_1 - c) - n_1 \log_2 N - c \log_2(n_1 - c) + c \log_2 N$$
$$+ c \log_2 1 - c \log_2 N.$$

Canceling terms yields

$$n_1 \log_2 n_1 > (n_1 - c) \log_2(n_1 - c),$$

which is obviously true, and it is proved.

Theorem 5.35. I_{Total} satisfies P3.

Proof. We need to show that

$$m \left(-\left(\left(\frac{n_1}{N} \right) \log_2 \left(\frac{n_1}{N} \right) + \sum_{i=2}^{m} \left(\frac{1}{N} \right) \log_2 \left(\frac{1}{N} \right) \right) \right) <$$
$$m \left(-\left(\left(\frac{n_1 - c}{N} \right) \log_2 \left(\frac{n_1 - c}{N} \right) + \sum_{i=2}^{m+c} \left(\frac{1}{N} \right) \log_2 \left(\frac{1}{N} \right) \right) \right).$$

Canceling terms and multiplying both sides of the above inequality by -1 yields

$$\left(\frac{n_1}{N} \right) \log_2 \left(\frac{n_1}{N} \right) + \sum_{i=2}^{m} \left(\frac{1}{N} \right) \log_2 \left(\frac{1}{N} \right) >$$
$$\left(\frac{n_1 - c}{N} \right) \log_2 \left(\frac{n_1 - c}{N} \right) + \sum_{i=2}^{m+c} \left(\frac{1}{N} \right) \log_2 \left(\frac{1}{N} \right),$$

which is proved in Theorem 5.34.

Theorem 5.36. I_{Max} satisfies P3.

Proof. We need to show that

$$\log_2 m < \log_2(m + c).$$

Now $c > 0$, so the above inequality is obviously true, and it is proved.

Theorem 5.37. $I_{McIntosh}$ satisfies P3.

Proof. We need to show that

$$\frac{N - \sqrt{n_1^2 + \sum_{i=2}^{m} 1^2}}{N - \sqrt{N}} < \frac{N - \sqrt{(n_1 - c)^2 + \sum_{i=2}^{m+c} 1^2}}{N - \sqrt{N}}.$$

Canceling terms and multiplying both sides of the above inequality by -1 yields

$$n_1^2 + m - 1 > (n_1 - c)^2 + (m + c - 1).$$

Canceling terms yields

$$0 > -2n_1 c + c^2 + c.$$

Moving $-2n_1 c$ to the left side of the above inequality and factoring c out of both sides yields

$$2n_1 > c + 1.$$

Now $c \geq 1$ and $n_1 > c$, so $2n_1 > c + 1$, and it is proved.

Theorem 5.38. I_{Lorenz} satisfies P3.

Proof. We need to show that

$$\left(\frac{1}{m}\right)\left(\frac{n_1 m}{N} + \sum_{i=2}^{m}(m-i+1)\left(\frac{1}{N}\right)\right) >$$
$$\left(\frac{1}{m+k}\right)\left(\frac{(n_1-k)m}{N} + \sum_{i=2}^{m+k}(m-i+1)\left(\frac{1}{N}\right)\right).$$

Since

$$\sum_{i=2}^{m}(m-i+1)\left(\frac{1}{N}\right) = \frac{(m+1)m}{N} + \frac{3(m+1)}{2N} - \frac{(m+1)^2}{2N} - \frac{2m}{N} - \frac{1}{N}$$

and

$$\sum_{i=2}^{m+k}(m-i+1)\left(\frac{1}{N}\right) =$$
$$\frac{(m+k+1)m}{N} + \frac{3(m+k+1)}{2N} - \frac{(m+k+1)^2}{2N} - \frac{2m}{N} - \frac{1}{N},$$

substituting for the summation terms in the above inequality yields

$$\left(\frac{1}{m}\right)\left(\frac{n_1 m}{N} + \frac{(m+1)m}{N} + \frac{3(m+1)}{2N} - \frac{(m+1)^2}{2N} - \frac{2m}{N} - \frac{1}{N}\right) >$$
$$\left(\frac{1}{m+k}\right)\left(\frac{(n_1-k)m}{N} + \frac{(m+k+1)m}{N} + \frac{3(m+k+1)}{2N}\right.$$
$$\left. - \frac{(m+k+1)^2}{2N} - \frac{2m}{N} - \frac{1}{N}\right).$$

Canceling terms yields

$$\frac{2n_1 k + 3mk + k^2}{2N(m+k)} > \frac{2k}{2N(m+k)}.$$

Factoring $k/(2N(m+k))$ out of both sides of the above inequality yields

$$2n_1 + 3m + k > 2.$$

Now $n_1 \geq 2$, $m > 1$, and $k > 0$, so the above inequality is obviously true, and it is proved.

Theorem 5.39. I_{Berger} satisfies P3.

Proof. We need to show that

$$\frac{n_1}{N} > \frac{n_1 - k}{N}.$$

Factoring $1/N$ out of both sides of the above inequality yields

$$n_1 > n_1 - k.$$

Canceling n_1 from both sides of the above inequality and moving k to the left side yields

$$k > 0,$$

and it is proved.

Theorem 5.40. $I_{Kullback}$ satisfies P3.

Proof. We need to show that

$$log_2 m - \left(\left(\frac{n_1}{N} \right) \log_2 \left(\frac{mn_1}{N} \right) + \sum_{i=2}^{m} \left(\frac{1}{N} \right) \log_2 \left(\frac{m}{N} \right) \right) <$$

$$log_2 m - \left(\left(\frac{n_1 - c}{N} \right) \log_2 \left(\frac{m(n_1 - c)}{N} \right) + \sum_{i=2}^{m+c} \left(\frac{1}{N} \right) \log_2 \left(\frac{m}{N} \right) \right).$$

Canceling terms and multiplying both sides of the above inequality by -1 yields

$$\left(\frac{n_1}{N} \right) \log_2 \left(\frac{mn_1}{N} \right) + \sum_{i=2}^{m} \left(\frac{1}{N} \right) \log_2 \left(\frac{m}{N} \right) >$$

$$\left(\frac{n_1 - c}{N} \right) \log_2 \left(\frac{m(n_1 - c)}{N} \right) + \sum_{i=2}^{m+c} \left(\frac{1}{N} \right) \log_2 \left(\frac{m}{N} \right).$$

Factoring $1/N$ out of both sides of the above inequality yields

$$n_1 \log_2 \left(\frac{mn_1}{N} \right) + (m - 1) \log_2 \left(\frac{m}{N} \right) >$$

$$(n_1 - c) \log_2 \left(\frac{m(n_1 - c)}{N} \right) + (m + c - 1) \log_2 \left(\frac{m}{N} \right).$$

Canceling terms yields

$$n_1 \log_2 \left(\frac{mn_1}{N} \right) > (n_1 - c) \log_2 \left(\frac{m(n_1 - c)}{N} \right) + c \log_2 \left(\frac{m}{N} \right).$$

Using the difference and addition properties of logarithms yields

$$n_1 \log_2 m + n_1 \log_2 n_1 - n_1 \log_2 N >$$
$$n_1 \log_2 m - c \log_2 m + (n_1 - c) \log_2 (n_1 - c) - n_1 \log_2 N + c \log_2 N$$
$$+ c \log_2 m - c \log_2 N.$$

Canceling terms yields

$$n_1 \log_2 n_1 > (n_1 - c) \log_2 (n_1 - c),$$

which is obviously true, and it is proved.

Theorem 5.41. I_{Lorenz}, I_{Gini}, I_{Schutz}, I_{Bray}, $I_{Whittaker}$, $I_{MacArthur}$, I_{Theil}, and $I_{Atkinson}$ do not satisfy P3.

Proof. Let $N = 4$, $m = 2$, $c = 1$, $n = (n_1, n_2) = (3, 1)$, and $n' = (n'_1, n'_2, n'_3) = (2, 1, 1)$.

5.3.4.   Permutation Invariance Principle

Permutation Invariance Principle (P4). Given a vector $(n_1, \ldots, n_m)$ and any permutation $(i_1, \ldots, i_m)$ of $(1, \ldots, m)$, $f(n_i, \ldots, n_m) = f(n_{i_1}, \ldots, n_{i_m})$.

Theorem 5.42. All measures except the I_{Lorenz} measure satisfy P4.

Proof. That this is true is immediately obvious due to the fact that the I_{Lorenz} measure not only uses the value of the summation index for referencing the values contained in the *Count* attribute, it also uses the value of the summation index in the calculation of the values that it generates. In contrast, the other measures use the value of the summation or multiplication index exclusively for referencing the values contained in the *Count* attribute.

5.3.5.   Transfer Principle

Transfer Principle (P5). Given a vector $(n_1, \ldots, n_m)$ and $0 < c < n_j$, $f(n_1, \ldots, n_i + c, \ldots, n_j - c, \ldots, n_m) > f(n_1, \ldots, n_i, \ldots, n_j, \ldots, n_m)$.

Theorem 5.43. $I_{Variance}$ satisfies P5.

Proof. We need to show that

$$\sum_{i=1}^{j-1} \left(\frac{n_i}{N} - \frac{1}{m} \right)^2 + \left(\frac{n_j + c}{N} - \frac{1}{m} \right)^2 + \sum_{i=j+1}^{k-1} \left(\frac{n_i}{N} - \frac{1}{m} \right)^2$$
$$+ \left(\frac{n_k - c}{N} - \frac{1}{m} \right)^2 + \sum_{i=k+1}^{m} \left(\frac{n_i}{N} - \frac{1}{m} \right)^2 >$$
$$\sum_{i=1}^{j-1} \left(\frac{n_i}{N} - \frac{1}{m} \right)^2 + \left(\frac{n_j}{N} - \frac{1}{m} \right)^2 + \sum_{i=j+1}^{k-1} \left(\frac{n_i}{N} - \frac{1}{m} \right)^2$$
$$+ \left(\frac{n_k}{N} - \frac{1}{m} \right)^2 + \sum_{i=k+1}^{m} \left(\frac{n_i}{N} - \frac{1}{m} \right)^2,$$

which after canceling terms is equivalent to showing that

$$\left(\frac{n_j + c}{N} - \frac{1}{m} \right)^2 + \left(\frac{n_k - c}{N} - \frac{1}{m} \right)^2 > \left(\frac{n_j}{N} - \frac{1}{m} \right)^2 + \left(\frac{n_k}{N} - \frac{1}{m} \right)^2.$$

Expanding both sides of the above inequality yields

$$\frac{n_j^2 + 2n_j c + 2c^2 + n_k^2 - 2n_k c}{N^2} - \frac{2(n_j + n_k)}{Nm} + \frac{2}{m^2} >$$
$$\frac{n_j^2 + n_k^2}{N^2} - \frac{2(n_j + n_k)}{Nm} + \frac{2}{m^2}.$$

Canceling terms yields

$$\frac{2(n_j c + c^2 - n_k c)}{N^2} > 0,$$

which is equivalent to

$$\frac{2(n_j c + c^2)}{N^2} > \frac{2n_k c}{N^2}.$$

Factoring $2c/N^2$ out of both sides of the above inequality yields

$$n_j + c > n_k.$$

Now $n_j > n_k$ and $c > 0$, so the above inequality is obviously true, and it is proved.

Theorem 5.44. $I_{Simpson}$ satisfies P5.

Proof. We need to show that

$$\sum_{i=1}^{j-1}\left(\frac{n_i}{N}\right)^2 + \left(\frac{n_j+c}{N}\right)^2 + \sum_{i=j+1}^{k-1}\left(\frac{n_i}{N}\right)^2 + \left(\frac{n_k-c}{N}\right)^2$$
$$+ \sum_{i=k+1}^{m}\left(\frac{n_i}{N}\right)^2 >$$
$$\sum_{i=1}^{j-1}\left(\frac{n_i}{N}\right)^2 + \left(\frac{n_j}{N}\right)^2 + \sum_{i=j+1}^{k-1}\left(\frac{n_i}{N}\right)^2 + \left(\frac{n_k}{N}\right)^2 + \sum_{i=k+1}^{m}\left(\frac{n_i}{N}\right)^2,$$

which after canceling terms is equivalent to showing that

$$\left(\frac{n_j+c}{N}\right)^2 + \left(\frac{n_k-c}{N}\right)^2 > \left(\frac{n_j}{N}\right)^2 + \left(\frac{n_k}{N}\right)^2.$$

Expanding both sides of the above inequality yields

$$\frac{n_j^2 + 2n_jc + c^2}{N^2} + \frac{n_k^2 + 2n_kc + c^2}{N^2} > \frac{n_j^2}{N^2} + \frac{n_k^2}{N^2}.$$

Canceling terms yields

$$\frac{2n_jc}{N^2} + \frac{2c^2}{N^2} - \frac{2n_kc}{N^2} > 0,$$

which is equivalent to

$$\frac{2n_jc}{N^2} + \frac{2c^2}{N^2} > \frac{2n_kc}{N^2}.$$

Factoring $2c/N^2$ out of both sides of the above inequality yields

$$n_j + c > n_k.$$

Now $n_j > n_k$ and $c > 0$, so the above inequality is obviously true, and it is proved.

Theorem 5.45. $I_{Shannon}$ satisfies P5.

Proof. We need to show that

$$-\left(\sum_{i=1}^{j-1}\left(\frac{n_i}{N}\right)\log_2\left(\frac{n_i}{N}\right) + \left(\frac{n_j}{N}\right)\log_2\left(\frac{n_j}{N}\right) + \sum_{i=j+1}^{k-1}\left(\frac{n_i}{N}\right)\log_2\left(\frac{n_i}{N}\right)\right.$$
$$\left. + \left(\frac{n_k}{N}\right)\log_2\left(\frac{n_k}{N}\right) + \sum_{i=k+1}^{m}\left(\frac{n_i}{N}\right)\log_2\left(\frac{n_i}{N}\right)\right) >$$

$$-\left(\sum_{i=1}^{j-1}\left(\frac{n_i}{N}\right)\log_2\left(\frac{n_i}{N}\right)+\left(\frac{n_j+c}{N}\right)\log_2\left(\frac{n_j+c}{N}\right)\right.$$

$$+\sum_{i=j+1}^{k-1}\left(\frac{n_i}{N}\right)\log_2\left(\frac{n_i}{N}\right)+\left(\frac{n_k-c}{N}\right)\log_2\left(\frac{n_k-c}{N}\right)$$

$$\left.+\sum_{i=k+1}^{m}\left(\frac{n_i}{N}\right)\log_2\left(\frac{n_i}{N}\right)\right).$$

Multiplying both sides of the above inequality by -1 and canceling terms yields

$$\left(\frac{n_j}{N}\right)\log_2\left(\frac{n_j}{N}\right)+\left(\frac{n_k}{N}\right)\log_2\left(\frac{n_k}{N}\right)<$$
$$\left(\frac{n_j+c}{N}\right)\log_2\left(\frac{n_j+c}{N}\right)+\left(\frac{n_k-c}{N}\right)\log_2\left(\frac{n_k-c}{N}\right).$$

Factoring $1/N$ out of both sides of the above inequality yields

$$n_j\log_2\left(\frac{n_j}{N}\right)+n_k\log_2\left(\frac{n_k}{N}\right)<$$
$$(n_j+c)\log_2\left(\frac{n_j+c}{N}\right)+(n_k-c)\log_2\left(\frac{n_k-c}{N}\right).$$

Using the difference property of logarithms to decompose the above inequality yields

$$n_j\log_2 n_j - n_j\log_2 N + n_k\log_2 n_k - n_k\log_2 N <$$
$$(n_j+c)\log_2(n_j+c)-(n_j+c)\log_2 N+(n_k-c)\log_2(n_k-c)$$
$$-(n_k-c)\log_2 N.$$

Canceling terms yields

$$n_j\log_2 n_j + n_k\log_2 n_k < (n_j+c)\log_2(n_j+c)+(n_k-c)\log_2(n_k-c).$$

Since the function $x\log_2 x$ is a convex function, we have

$$(n_j+c)\log_2(n_j+c)+(n_k-c)\log_2(n_k-c)\geq$$
$$(n_j+c+n_k-c)\log_2(n_j+c+n_k-c),$$

which is equivalent to

$$(n_j+c)\log_2(n_j+c)+(n_k-c)\log_2(n_k-c)\geq(n_j+n_k)\log_2(n_j+n_k).$$

We now show that

$$n_j\log_2 n_j + n_k\log_2 n_k < (n_j+n_k)\log_2(n_j+n_k).$$

Distributing $(n_j + n_k)$ over the log on the right side of the above inequality yields

$$n_j \log_2 n_j + n_k \log_2 n_k < n_j \log_2(n_j + n_k) + n_k \log_2(n_j + n_k),$$

which is obviously true, and it is proved.

Theorem 5.46. I_{Total} satisfies P5.

Proof. We need to show that

$$m \left(- \left(\sum_{i=1}^{j-1} \left(\frac{n_i}{N} \right) \log_2 \left(\frac{n_i}{N} \right) + \left(\frac{n_j}{N} \right) \log_2 \left(\frac{n_j}{N} \right) \right. \right.$$

$$+ \sum_{i=j+1}^{k-1} \left(\frac{n_i}{N} \right) \log_2 \left(\frac{n_i}{N} \right) + \left(\frac{n_k}{N} \right) \log_2 \left(\frac{n_k}{N} \right)$$

$$\left. \left. + \sum_{i=k+1}^{m} \left(\frac{n_i}{N} \right) \log_2 \left(\frac{n_i}{N} \right) \right) \right) >$$

$$m \left(- \left(\sum_{i=1}^{j-1} \left(\frac{n_i}{N} \right) \log_2 \left(\frac{n_i}{N} \right) + \left(\frac{n_j + c}{N} \right) \log_2 \left(\frac{n_j + c}{N} \right) \right. \right.$$

$$+ \sum_{i=j+1}^{k-1} \left(\frac{n_i}{N} \right) \log_2 \left(\frac{n_i}{N} \right) + \left(\frac{n_k - c}{N} \right) \log_2 \left(\frac{n_k - c}{N} \right)$$

$$\left. \left. + \sum_{i=k+1}^{m} \left(\frac{n_i}{N} \right) \log_2 \left(\frac{n_i}{N} \right) \right) \right).$$

Canceling terms and multiplying both sides of the above inequality by -1 yields

$$\sum_{i=1}^{j-1} \left(\frac{n_i}{N} \right) \log_2 \left(\frac{n_i}{N} \right) + \left(\frac{n_j}{N} \right) \log_2 \left(\frac{n_j}{N} \right) + \sum_{i=j+1}^{k-1} \left(\frac{n_i}{N} \right) \log_2 \left(\frac{n_i}{N} \right)$$

$$+ \left(\frac{n_k}{N} \right) \log_2 \left(\frac{n_k}{N} \right) + \sum_{i=k+1}^{m} \left(\frac{n_i}{N} \right) \log_2 \left(\frac{n_i}{N} \right) <$$

$$\sum_{i=1}^{j-1} \left(\frac{n_i}{N} \right) \log_2 \left(\frac{n_i}{N} \right) + \left(\frac{n_j + c}{N} \right) \log_2 \left(\frac{n_j + c}{N} \right)$$

$$+ \sum_{i=j+1}^{k-1} \left(\frac{n_i}{N} \right) \log_2 \left(\frac{n_i}{N} \right) + \left(\frac{n_k - c}{N} \right) \log_2 \left(\frac{n_k - c}{N} \right)$$

$$+ \sum_{i=k+1}^{m} \left(\frac{n_i}{N} \right) \log_2 \left(\frac{n_i}{N} \right),$$

which is proved in Theorem 5.45.

Theorem 5.47. $I_{McIntosh}$ satisfies P5.

Proof. We need to show that

$$\frac{N - \sqrt{\sum_{i=1}^{j-1} n_i^2 + (n_j + c)^2 + \sum_{i=j+1}^{k-1} n_i^2 + (n_k - c)^2 + \sum_{i=k+1}^{m} n_i^2}}{N - \sqrt{N}} <$$

$$\frac{N - \sqrt{\sum_{i=1}^{j-1} n_i^2 + n_j^2 + \sum_{i=j+1}^{k-1} n_i^2 + n_k^2 + \sum_{i=k+1}^{m} n_i^2}}{N - \sqrt{N}}.$$

Canceling terms and multiplying both sides of the above inequality by -1 yields

$$(n_j + c)^2 + (n_k - c)^2 > n_j^2 + n_k^2.$$

Expanding the terms in the left side yields

$$n_j^2 + 2n_j c + c^2 + n_k^2 - 2n_k c + c^2 > n_j^2 + n_k^2.$$

Canceling terms yields and moving $-2n_k c$ to the right side yields

$$2n_j c + 2c^2 > 2n_k c.$$

Factoring $2c$ out of both sides of the above inequality yields

$$n_j + c > n_k.$$

Now $n_j > n_k$ and $c > 0$, so the above inequality is obviously true, and it is proved.

Theorem 5.48. I_{Lorenz} satisfies P5.

Proof. We need to show that

$$\frac{1}{m} \left(\sum_{i=1}^{j-1} (m - i + 1) \left(\frac{n_i}{N} \right) + (m - j + 1) \left(\frac{n_j}{N} \right) \right.$$

$$+ \sum_{i=j+1}^{k-1} (m - i + 1) \left(\frac{n_i}{N} \right) + (m - k + 1) \left(\frac{n_k}{N} \right)$$

$$\left. + \sum_{i=k+1}^{m} (m - i + 1) \left(\frac{n_i}{N} \right) \right) >$$

$$\frac{1}{m} \left(\sum_{i=1}^{j-1} (m-i+1) \left(\frac{n_i}{N} \right) + (m-j+1) \left(\frac{n_j - c}{N} \right) \right.$$

$$+ \sum_{i=j+1}^{k-1} (m-i+1) \left(\frac{n_i}{N} \right) + (m-k+1) \left(\frac{n_k + c}{N} \right)$$

$$\left. + \sum_{i=k+1}^{m} (m-i+1) \left(\frac{n_i}{N} \right) \right).$$

Canceling terms yields

$$(m-j+1) \left(\frac{n_j}{N} \right) + (m-k+1) \left(\frac{n_k}{N} \right) >$$

$$(m-j+1) \left(\frac{n_j - c}{N} \right) + (m-k+1) \left(\frac{n_k + c}{N} \right).$$

Expanding terms on the right side of the above inequality yields

$$(m-j+1) \left(\frac{n_j}{N} \right) + (m-k+1) \left(\frac{n_k}{N} \right) >$$

$$(m-j+1) \left(\frac{n_j}{N} - (m-j+1)\frac{c}{N} \right)$$

$$+ (m-k+1) \left(\frac{n_k}{N} + (m-k+1)\frac{c}{N} \right).$$

Canceling terms yields

$$0 > -(m-j+1) \left(\frac{c}{N} \right) + (m-k+1) \left(\frac{c}{N} \right).$$

Moving $-(m-j+1) \left(\frac{c}{N} \right)$ to the left side yields

$$(m-j+1) \left(\frac{c}{N} \right) > (m-k+1) \left(\frac{c}{N} \right).$$

Factoring c/N out of both sides of the above inequality and canceling terms yields

$$-j > -k$$

which is equivalent to

$$j < k,$$

which is obviously true, and it is proved.

Theorem 5.49. I_{Gini} satisfies P5.

Proof. To simplify the proof of this theorem, we again use the equivalent form of the Gini index introduced in the proof of Theorem 5.23. We need to show that

$$2\left(\frac{m+1}{2} - \left(\sum_{i=1}^{j-1} i\left(\frac{n_i}{N}\right) + j\left(\frac{n_j+c}{N}\right) + \sum_{i=j+1}^{k-1} i\left(\frac{n_i}{N}\right)\right.\right.$$
$$\left.\left. +k\left(\frac{n_k-c}{N}\right) + \sum_{i=k+1}^{m} i\left(\frac{n_i}{N}\right)\right)\right)/m >$$

$$2\left(\frac{m+1}{2} - \left(\sum_{i=1}^{j-1} i\left(\frac{n_i}{N}\right) + j\left(\frac{n_j}{N}\right) + \sum_{i=j+1}^{k-1} i\left(\frac{n_i}{N}\right)\right.\right.$$
$$\left.\left. +k\left(\frac{n_k}{N}\right) + \sum_{i=k+1}^{m} i\left(\frac{n_i}{N}\right)\right)\right)/m.$$

Canceling terms from both sides of the above inequality yields

$$-j\left(\frac{n_j+c}{N}\right) - k\left(\frac{n_k-c}{N}\right) > -j\left(\frac{n_j}{N}\right) - k\left(\frac{n_k}{N}\right),$$

which is equivalent to

$$-\frac{jc}{N} + \frac{kc}{N} > 0.$$

Factoring c/N out of the left side of the above inequality yields and moving $-j$ to the right side yields

$$k > j,$$

which is obviously true, and it is proved.

Theorem 5.50. $I_{Kullback}$ satisfies P5.

Proof. We need to show that

$$log_2 m - \left(\sum_{i=1}^{j-1} \left(\frac{n_i}{N}\right) log_2\left(\frac{mn_i}{N}\right) + \left(\frac{n_j}{N}\right) log_2\left(\frac{mn_j}{N}\right)\right.$$

$$+ \sum_{i=j+1}^{k-1} \left(\frac{n_i}{N}\right) log_2\left(\frac{mn_i}{N}\right) + \left(\frac{n_k}{N}\right) log_2\left(\frac{mn_k}{N}\right)$$

$$\left. + \sum_{i=k+1}^{m} \left(\frac{n_i}{N}\right) log_2\left(\frac{mn_i}{N}\right)\right) >$$

$$log_2 m - \left(\sum_{i=1}^{j-1} \left(\frac{n_i}{N} \right) \log_2 \left(\frac{mn_i}{N} \right) + \left(\frac{n_j + c}{N} \right) \log_2 \left(\frac{m(n_j + c)}{N} \right) \right.$$

$$+ \sum_{i=j+1}^{k-1} \left(\frac{n_i}{N} \right) \log_2 \left(\frac{mn_i}{N} \right) + \left(\frac{n_k - c}{N} \right) \log_2 \left(\frac{m(n_k - c)}{N} \right)$$

$$\left. + \sum_{i=k+1}^{m} \left(\frac{n_i}{N} \right) \log_2 \left(\frac{mn_i}{N} \right) \right).$$

Multiplying both sides of the above inequality by -1 and canceling terms yields

$$\left(\frac{n_j}{N} \right) \log_2 \left(\frac{mn_j}{N} \right) + \left(\frac{n_k}{N} \right) \log_2 \left(\frac{mn_k}{N} \right) <$$
$$\left(\frac{n_j + c}{N} \right) \log_2 \left(\frac{m(n_j + c)}{N} \right) + \left(\frac{n_k - c}{N} \right) \log_2 \left(\frac{m(n_k - c)}{N} \right).$$

Factoring $1/N$ out of both sides of the above inequality yields

$$n_j \log_2 \left(\frac{mn_j}{N} \right) + n_k \log_2 \left(\frac{mn_k}{N} \right) <$$
$$(n_j + c) \log_2 \left(\frac{m(n_j + c)}{N} \right) + (n_k - c) \log_2 \left(\frac{m(n_k - c)}{N} \right).$$

Using the difference and addition properties of logarithms to decompose the above inequality yields

$$n_j \log_2 m + n_j \log_2 n_j - n_j \log_2 N + n_k \log_2 m + n_k \log_2 n_k$$
$$- n_k \log_2 N <$$
$$n_j \log_2 m + c \log_2 m + (n_j + c) \log_2 (n_j + c) - n_j \log_2 N$$
$$- c \log_2 N + n_k \log_2 m - c \log_2 m + (n_k - c) \log_2 (n_k - c)$$
$$- n_k \log_2 N + c \log_2 N.$$

Canceling terms yields

$$n_j \log_2 n_j + n_k \log_2 n_k < (n_j + c) \log_2 (n_j + c) + (n_k - c) \log_2 (n_k - c),$$

which is proved in Theorem 5.45.

Theorem 5.51. $I_{MacArthur}$ satisfies P5.

Proof. Let the vector $(a, \ldots, a)$ containing m a's be the uniform distribution. Then the distribution obtained by combining the vectors $(n_1 + c, n_2 - d_2, \ldots, n_m - d_m)$ and $(a, \ldots, a)$ is the vector $(n_1 + c + a, n_2 - d_2 + a, \ldots, n_m -$

$d_m + a$), and the distribution obtained by combining the vectors $(n_1, \ldots, n_m)$ and $(a, \ldots, a)$ is the vector $(n_1 + a, \ldots, n_m + a)$. We need to show that

$$
-\left(\sum_{i=1}^{j-1} \left(\frac{n_i + a}{2N} \right) \log_2 \left(\frac{n_i + a}{2N} \right) + \left(\frac{n_j + c + a}{2N} \right) \log_2 \left(\frac{n_j + c + a}{2N} \right) \right.
$$

$$
+ \sum_{i=j+1}^{k-1} \left(\frac{n_i + a}{2N} \right) \log_2 \left(\frac{n_i + a}{2N} \right) + \left(\frac{n_k - c + a}{2N} \right) \log_2 \left(\frac{n_k - c + a}{2N} \right)
$$

$$
+ \sum_{i=k+1}^{m} \left(\frac{n_i + a}{2N} \right) \log_2 \left(\frac{n_i + a}{2N} \right) \right) - \left(- \left(\sum_{i=1}^{j-1} \left(\frac{n_i}{N} \right) \log_2 \left(\frac{n_i}{N} \right) \right. \right.
$$

$$
+ \left(\frac{n_j + c}{N} \right) \log_2 \left(\frac{n_j + c}{N} \right) + \sum_{i=j+1}^{k-1} \left(\frac{n_i}{N} \right) \log_2 \left(\frac{n_i}{N} \right)
$$

$$
+ \left(\frac{n_k - c}{N} \right) \log_2 \left(\frac{n_k - c}{N} \right) + \sum_{i=k+1}^{m} \left(\frac{n_i}{N} \right) \log_2 \left(\frac{n_i}{N} \right) \right)
$$

$$
\left. + \log_2 m \right) \bigg/ 2 >
$$

$$
-\left(\sum_{i=1}^{j-1} \left(\frac{n_i + a}{2N} \right) \log_2 \left(\frac{n_i + a}{2N} \right) + \left(\frac{n_j + a}{2N} \right) \log_2 \left(\frac{n_j + a}{2N} \right) \right.
$$

$$
+ \sum_{i=j+1}^{k-1} \left(\frac{n_i + a}{2N} \right) \log_2 \left(\frac{n_i + a}{2N} \right) + \left(\frac{n_k + a}{2N} \right) \log_2 \left(\frac{n_k + a}{2N} \right)
$$

$$
+ \sum_{i=k+1}^{m} \left(\frac{n_i + a}{2N} \right) \log_2 \left(\frac{n_i + a}{2N} \right) \right) - \left(- \left(\sum_{i=1}^{j-1} \left(\frac{n_i}{N} \right) \log_2 \left(\frac{n_i}{N} \right) \right. \right.
$$

$$
+ \left(\frac{n_j}{N} \right) \log_2 \left(\frac{n_j}{N} \right) + \sum_{i=j+1}^{k-1} \left(\frac{n_i}{N} \right) \log_2 \left(\frac{n_i}{N} \right)
$$

$$
+ \left(\frac{n_k}{N} \right) \log_2 \left(\frac{n_k}{N} \right) + \sum_{i=k+1}^{m} \left(\frac{n_i}{N} \right) \log_2 \left(\frac{n_i}{N} \right) \right)
$$

$$
\left. + \log_2 m \right) \bigg/ 2,
$$

which is equivalent to

$$-\sum_{i=1}^{j-1} \left(\frac{n_i + a}{2N}\right) \log_2 \left(\frac{n_i + a}{2N}\right) - \left(\frac{n_j + c + a}{2N}\right) \log_2 \left(\frac{n_j + c + a}{2N}\right)$$

$$-\sum_{i=j+1}^{k-1} \left(\frac{n_i + a}{2N}\right) \log_2 \left(\frac{n_i + a}{2N}\right) - \left(\frac{n_k - c + a}{2N}\right) \log_2 \left(\frac{n_k - c + a}{2N}\right)$$

$$-\sum_{i=k+1}^{m} \left(\frac{n_i + a}{2N}\right) \log_2 \left(\frac{n_i + a}{2N}\right) + \sum_{i=1}^{j-1} \left(\frac{n_i}{2N}\right) \log_2 \left(\frac{n_i}{N}\right)$$

$$+ \left(\frac{n_j + c}{2N}\right) \log_2 \left(\frac{n_j + c}{N}\right) + \sum_{i=j+1}^{k-1} \left(\frac{n_i}{2N}\right) \log_2 \left(\frac{n_i}{N}\right)$$

$$+ \left(\frac{n_k - c}{2N}\right) \log_2 \left(\frac{n_k - c}{N}\right) + \sum_{i=k+1}^{m} \left(\frac{n_i}{2N}\right) \log_2 \left(\frac{n_i}{N}\right) - \frac{\log_2 m}{2} >$$

$$-\sum_{i=1}^{j-1} \left(\frac{n_i + a}{2N}\right) \log_2 \left(\frac{n_i + a}{2N}\right) - \left(\frac{n_j + a}{2N}\right) \log_2 \left(\frac{n_j + a}{2N}\right)$$

$$-\sum_{i=j+1}^{k-1} \left(\frac{n_i + a}{2N}\right) \log_2 \left(\frac{n_i + a}{2N}\right) - \left(\frac{n_k + a}{2N}\right) \log_2 \left(\frac{n_k + a}{2N}\right)$$

$$-\sum_{i=k+1}^{m} \left(\frac{n_i + a}{2N}\right) \log_2 \left(\frac{n_i + a}{2N}\right) + \sum_{i=1}^{j-1} \left(\frac{n_i}{2N}\right) \log_2 \left(\frac{n_i}{N}\right)$$

$$+ \left(\frac{n_j}{2N}\right) \log_2 \left(\frac{n_j}{N}\right) + \sum_{i=j+1}^{k-1} \left(\frac{n_i}{2N}\right) \log_2 \left(\frac{n_i}{N}\right)$$

$$+ \left(\frac{n_k}{2N}\right) \log_2 \left(\frac{n_k}{N}\right) + \sum_{i=k+1}^{m} \left(\frac{n_i}{2N}\right) \log_2 \left(\frac{n_i}{N}\right) - \frac{\log_2 m}{2}.$$

Canceling terms, factoring $1/2N$ out of both sides of the above inequality, and moving terms preceded by a minus sign to the other side of the inequality yields

$$(n_j + c) \log_2 \left(\frac{n_j + c}{N}\right) + (n_k - c) \log_2 \left(\frac{n_k - c}{N}\right)$$

$$+ (n_j + a) \log_2 \left(\frac{n_j + a}{2N}\right) + (n_k + a) \log_2 \left(\frac{n_k + a}{2N}\right) >$$

$$(n_j + c + a) \log_2 \left(\frac{n_j + c + a}{2N}\right) + (n_k - c + a) \log_2 \left(\frac{n_k - c + a}{2N}\right)$$

$$+ n_j \log_2 \left(\frac{n_j}{N}\right) + n_k \log_2 \left(\frac{n_k}{N}\right).$$

Using the difference property of logarithms to decompose the above inequality yields

$$
\begin{aligned}
&(n_j + c) \log_2(n_j + c) - n_j \log_2 N - c \log_2 N + (n_k - c) \log_2(n_k - c) \\
&- n_k \log_2 N + c \log_2 N + (n_j + a) \log_2(n_j + a) - n_j \log_2 2N \\
&- a \log_2 2N + (n_k + a) \log_2(n_k + a) - n_k \log_2 2N - a \log_2 2N > \\
&(n_j + c + a) \log_2(n_j + c + a) - n_j \log_2 2N - c \log_2 2N \\
&- a \log_2 2N + (n_k - c + a) \log_2(n_k - c + a) - n_k \log_2 2N \\
&+ c \log_2 2N - a \log_2 2N + n_j \log_2 n_j - n_j \log_2 N + n_k \log_2 n_k \\
&- n_k \log_2 N.
\end{aligned}
$$

Canceling terms yields

$$
\begin{aligned}
&(n_j + c) \log_2(n_j + c) + (n_k - c) \log_2(n_k - c) + (n_j + a) \log_2(n_j + a) \\
&+ (n_k + a) \log_2(n_k + a) > \\
&(n_j + c + a) \log_2(n_j + c + a) + (n_k - c + a) \log_2(n_k - c + a) \\
&+ n_j \log_2 n_j + n_k \log_2 n_k.
\end{aligned}
$$

To show that the above inequality is true, we need to show that

$$
(n_j + c) \log_2(n_j + c) + (n_j + a) \log_2(n_j + a) >
$$

$$
(n_j + c + a) \log_2(n_j + c + a) + n_j \log_2 n_j
$$

and

$$
(n_k + a) \log_2(n_k + a) + (n_k - c) \log_2(n_k - c) >
$$

$$
n_k \log_2 n_k + (n_k - c + a) \log_2(n_k - c + a).
$$

We now show that

$$
(n_j + c) \log_2(n_j + c) + (n_j + a) \log_2(n_j + a) >
$$

$$
(n_j + c + a) \log_2(n_j + c + a) + n_j \log_2 n_j.
$$

Since the function $x \log_2 x$ is a convex function, we have

$$
\begin{aligned}
&(n_j + c) \log_2(n_j + c) + (n_j + a) \log_2(n_j + a) \geq \\
&(n_j + c + n_j + a) \log_2(n_j + c + n_j + a),
\end{aligned}
$$

which is equivalent to

$$
\begin{aligned}
&(n_j + c) \log_2(n_j + c) + (n_j + a) \log_2(n_j + a) \geq \\
&n_j \log_2(n_j + c + n_j + a) + (n_j + c + a) \log_2(n_j + c + n_j + a),
\end{aligned}
$$

so obviously

$$(n_j + c) \log_2(n_j + c) + (n_j + a) \log_2(n_j + a) >$$
$$n_j \log_2 n_j + (n_j + c + a) \log_2(n_j + c + a),$$

and the first part is shown. We now show that

$$(n_k + a) \log_2(n_k + a) + (n_k - c) \log_2(n_k - c) >$$
$$n_k \log_2 n_k + (n_k - c + a) \log_2(n_k - c + a).$$

Since the function $x \log_2 x$ is a convex function, we have

$$(n_k + a) \log_2(n_k + a) + (n_k - c) \log_2(n_k - c) \geq$$
$$(n_k + a + n_k - c) \log_2(n_k + a + n_k - c),$$

which is equivalent to

$$(n_j + c) \log_2(n_j + c) + (n_j + a) \log_2(n_j + a) \geq$$
$$n_k \log_2(n_k + a + n_k - c) + (n_k - c + a) \log_2(n_k + a + n_k - c),$$

so obviously

$$(n_j + c) \log_2(n_j + c) + (n_j + a) \log_2(n_j + a) >$$
$$n_k \log_2 n_k + (n_k - c + a) \log_2(n_k - c + a),$$

the second part is shown, and it is proved.

Theorem 5.52. $I_{Atkinson}$ satisfies P5.

Proof. We need to show that

$$1 - \left(\prod_{i=1}^{j-1} \frac{\left(\frac{n_i}{N}\right)}{\left(\frac{1}{m}\right)} \frac{\left(\frac{n_j+c}{N}\right)}{\left(\frac{1}{m}\right)} \prod_{i=j+1}^{k-1} \frac{\left(\frac{n_i}{N}\right)}{\left(\frac{1}{m}\right)} \frac{\left(\frac{n_k-c}{N}\right)}{\left(\frac{1}{m}\right)} \prod_{i=k+1}^{m} \frac{\left(\frac{n_i}{N}\right)}{\left(\frac{1}{m}\right)} \right)^{\frac{1}{m}} >$$

$$1 - \left(\prod_{i=1}^{j-1} \frac{\left(\frac{n_i}{N}\right)}{\left(\frac{1}{m}\right)} \frac{\left(\frac{n_j}{N}\right)}{\left(\frac{1}{m}\right)} \prod_{i=j+1}^{k-1} \frac{\left(\frac{n_i}{N}\right)}{\left(\frac{1}{m}\right)} \frac{\left(\frac{n_k}{N}\right)}{\left(\frac{1}{m}\right)} \prod_{i=k+1}^{m} \frac{\left(\frac{n_i}{N}\right)}{\left(\frac{1}{m}\right)} \right)^{\frac{1}{m}}.$$

Canceling terms from both sides of the above inequality yields

$$-((n_j + c)(n_k - c)) > -n_j n_k,$$

Expanding the left side of the above inequality yields

$$-n_j n_k + n_j c - n_k c + c^2 > -n_j n_k.$$

Canceling terms and moving $-n_k c$ to the right side of the above inequality yields

$$n_j c + c^2 > n_k c.$$

Factoring c out of both sides of the above inequality yields

$$n_j + c > n_k.$$

Now $n_j > n_k$ and $c > 0$, so the above inequality is obviously true, and it is proved.

Theorem 5.53. I_{Max}, I_{Berger}, I_{Schutz}, I_{Bray}, $I_{Whittaker}$, and I_{Theil} do not satisfy P5.

Proof. Let $N = 7$, $m = 3$, $n = (n_1, n_2, n_3) = (4, 2, 1)$, and $n' = (n_1', n_2', n_3') = (3, 2, 2)$.

Chapter 6

EXPERIMENTAL ANALYSES

As mentioned in Chapter 3, our data mining algorithm, *All_Gen*, has been implemented in the research software tool *DGG-Discover*. In this chapter, we evaluate the performance of *All_Gen* in generating summaries from databases. We also evaluate the sixteen diversity measures for ranking the interestingness of the summaries generated, implemented in the research software tool *DGG-Interest*. We present the results of our evaluation against a variety of metrics and describe our general experience.

6.1. Evaluation of the *All_Gen* Algorithm

DGG-Discover was developed to evaluate the serial and parallel versions of the *All_Gen* algorithm. We ran all of our experiments on a 64-node Alex AVX Series 2, a MIMD distributed-memory parallel computer. Each inside-the-box compute node consists of a T805 processor, with 8 MB of local memory, paired with an i860 processor, with 32 MB of shared memory (the pair communicates through the shared memory). Each i860 processor runs at 40 MHz and each T805 processor runs at 20 MHz with a bandwidth of 20 Mbits/second of bidirectional data throughput on each of its four links. The compute nodes run version 2.2.3 of the Alex-Trollius operating system. The front-end host computer system, a Sun Sparc 20 with 32 MB of memory, runs version 2.4 of the Solaris operating system and uses Oracle Release 7.3 for database management.

DGG-Discover functions as three types of communicating modules: a *slave program* runs on an inside-the-box compute node and executes the discovery tasks that it is assigned, the *master program* assigns discovery tasks to the slave programs, and the *bridge program* coordinates access between the slave programs and the database.

Input data was from the Customer Database, a confidential database supplied by a commercial research partner in the telecommunications industry. This database has been used frequently in previous data mining research [26, 63, 72, 64]. It consists of over 8,000,000 tuples in 22 tables describing a total of 56 attributes. The largest table contains over 3,300,000 tuples representing the account activity for over 500,000 customer accounts and over 2,200 products and services. Our queries read approximately 545,000 tuples from three tables, resulting in an initial input relation for the discovery tasks containing up to 26,950 tuples and five attributes. Our experience in applying data mining techniques to the databases of our commercial research partners has shown that domain experts typically perform discovery tasks on a few attributes that have been determined to be relevant. Consequently, we present the results for experiments where two to five attributes are selected for generalization, and the DGGs associated with the selected attributes contained one to five unique paths. The characteristics of the DGGs associated with each attribute are shown in Table 6.1, where the *No. of Paths* column describes the number of unique paths, the *No. of Nodes* column describes the number of nodes, and the *Avg. Path Length* column describes the average path length.

Table 6.1.　Characteristics of the DGGs associated with the selected attributes

Attribute	No. of Paths	No. of Nodes	Avg. Path Length
A	5	20	4.0
B	4	17	4.3
C	3	12	4.0
D	4	17	4.3
E	2	8	4.0
F	1	3	3.0
G	5	21	4.2

From these experiments, we draw two main conclusions.

- As the complexity of the DGGs associated with a discovery task increases (either by adding more paths to a DGG, more nodes to a path, or more attributes to a discovery task), the complexity and traversal time of the generalization space also increases.

- As the number of processors used in a discovery task increases, the time required to traverse the generalization space decreases, resulting in significant speedups for discovery tasks run on multiple processors.

6.1.1.　Serial vs Parallel Performance

We now compare the serial and parallel performance of the indexAll_Gen algorithm on a variety of discovery tasks. The results are shown in the graphs of

Figures 6.1 to 6.4, where the number of processors is plotted against execution time in seconds. In each of the four experiments discussed here, we varied the number of processors assigned to the discovery tasks. A maximum of 32 processors were available. The graphs show that as the complexity of the generalization space increases, the time required to traverse the generalization space also increases. For example, the two-, three-, four-, and five-attribute discovery tasks in Figures 6.1, 6.2, 6.3, and 6.4, respectively, required a serial time of 36, 402, 3,732, and 25,787 seconds, respectively, to generate 340, 3,468, 27,744, and 102,816 summaries, respectively, on a single processor. A similar result was obtained when multiple processors were allocated to each discovery task. The number of summaries to be generated by a discovery task (i.e., the size of the generalization space) is determined by multiplying the values in the *No. of Nodes* column of Table 6.1. For example, when attributes B, C, D, and E were selected for the four-attribute discovery task, 27,744 (i.e., $17 \times 12 \times 17 \times 8$) summaries were generated.

The graphs also show that as the number of processors assigned to a discovery task is increased, the time required to traverse the generalization space decreases. Each discovery task can be divide into smaller discovery tasks (i.e., sub-tasks) which can each be run independently on a separate processor. For example, the two-, three-, four-, and five-attribute discovery tasks that required 36, 402, 3,732, and 25,787 seconds, respectively, on a single processor, required

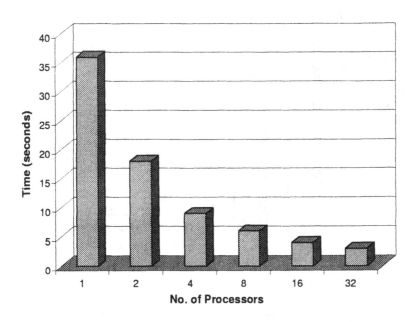

Figure 6.1. Relative performance generalizing two attributes

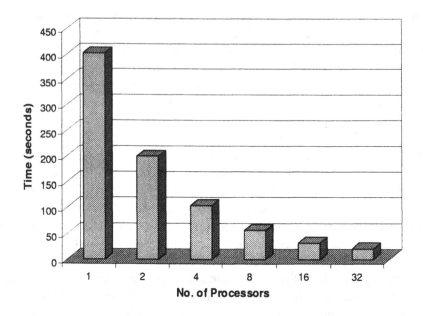

Figure 6.2.   Relative performance generalizing three attributes

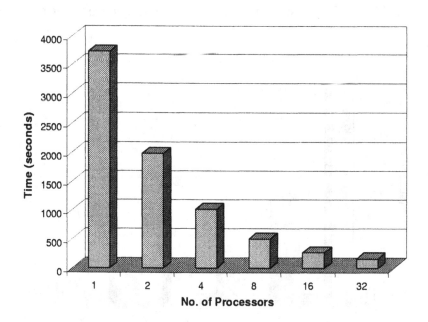

Figure 6.3.   Relative performance generalizing four attributes

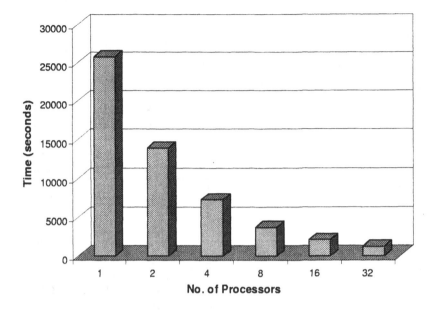

Figure 6.4.   Relative performance generalizing five attributes

only 3, 21, 167, and 1,245 seconds, respectively, on 32 processors. The two-attribute discovery task was partitioned across 20 of the 32 available processors, as there were only 20 possible sub-tasks (i.e., unique path combinations). The number of sub-tasks to be generated by a discovery task is determined by multiplying the values in the *No. of Paths* column of Table 6.1. For example, when attributes B, C, D, and E were selected for the four-attribute discovery task, the discovery task could be partitioned into 96 (i.e., $4 \times 3 \times 4 \times 2$) sub-tasks.

6.1.2.   Speedup and Efficiency Improvements

Speedups for the discovery tasks run on multiple processors are shown in Table 6.2. In Table 6.2, the *No. of Nodes* column describes the number of nodes in the generalization space, the *No. of Sub-Tasks* column describes the number of unique path combinations that can be obtained from the set of DGGs associated with the attributes, the *No. of Processors* column describes the number of processors used, the *Time* column describes the time required to run the discovery task on the corresponding number of processors, the *Speedup* column describes the serial time (i.e., the time required on one processor) divided by the parallel time (i.e., the time required on multiple processors), and the *Efficiency* column describes the speedup divided by the number of processors. Significant speedups were obtained when a discovery task was run on multiple processors. For example, the maximum speedups for the two-,

Table 6.2.   Speedup and efficiency results obtained using the parallel algorithm

Attributes	No. of Nodes	No. of Sub-Tasks	No. of Processors	Time	Speedup	Efficiency
A,B	340	20	1	36	-	-
			2	18	2.0	1.00
			4	9	4.0	1.00
			8	6	6.0	0.75
			16	4	9.0	0.56
			20	3	12.0	0.60
B,C,D	3468	48	1	402	-	-
			2	199	2.0	1.00
			4	104	3.9	0.98
			8	56	7.2	0.90
			16	32	12.6	0.79
			32	21	19.1	0.60
B,C,D,E	27744	96	1	3732	-	-
			2	1985	1.9	0.95
			4	1017	3.7	0.93
			8	506	7.4	0.93
			16	273	13.7	0.86
			32	167	22.3	0.70
C,D,E,F,G	102816	120	1	25787	-	-
			2	13939	1.8	0.90
			4	7264	3.5	0.88
			8	3723	6.9	0.86
			16	2080	12.4	0.78
			32	1245	20.7	0.65

three-, four-, and five-attribute discovery tasks, which were obtained when the discovery tasks were run on 32 processors, were 12.0, 19.1, 22.3, and 20.7, respectively.

6.2.   Evaluation of the Sixteen Diversity Measures

DGG-Interest was developed to evaluate the sixteen diversity measures when used for ranking the interestingness of summaries. We ran all of our experiments on a 12-node Silicon Graphics Challenge M, a SMP shared-memory parallel computer. Each 150 MHz IP19 processor consists of a MIPS R4400 CPU paired with a MIPS R4010 FPU. There are 512 MB of shared memory. The computer system runs Release 5.3 of the IRIX operating system and uses Oracle Release 7.3 for database management.

Input data was from the Research Awards Database, a database supplied by NSERC and available in the public domain, and the Customer Database described in the previous section. The Research Awards Database has also been frequently used in previous data mining research [22, 24, 49, 90]. It consists of 10,000 tuples in six tables describing a total of 22 attributes.

We again present results for experiments where two to five attributes are selected for generalization. Discovery tasks were run against the NSERC database, from which two to four attributes were selected for discovery, and against the Customer database, from which two to five attributes were selected for discovery. We refer to the NSERC discovery tasks containing two, three, and four attributes as as *N-2*, *N-3*, and *N-4*, respectively, and the Customer discovery tasks containing two, three, four, and five attributes as *C-2*, *C-3*, *C-4*, and *C-5*, respectively.

From these experiments, we draw seven main conclusions.

- There are numerous ties in the ranks assigned by each measure.

- The ranks assigned by some measures have a high positive correlation.

- There are two distinct groups of measures within which summaries are ranked similarly. While it might be expected that the measures taken from a particular discipline would constitute a group, this is not necessarily the case.

- While there were combinations of measures that showed no correlation, there were no highly negative correlations.

- The summaries ranked as most interesting tend to be more concise, containing few tuples and/or non-ANY attributes, although there were a few minor exceptions.

- The index values generated by most of the measures tend to be highly skewed about the mean, middle, and median index values.

- The vectors associated with the middle index values generated by most of the measures are too skewed.

6.2.1.   Comparison of Assigned Ranks

We now compare the ranks assigned to the summaries by each measure. Typical results are shown in Table 6.3 and Tables A.1 through A.7 of Appendix A, where 22 summaries, generated from the *N-2* discovery task, are ranked by the various measures. In Table 6.3 and Tables A.1 through A.7, the *Summary ID* column describes a unique summary identifier (for reference purposes), the *Non-ANY Attributes* column describes the number of attributes that have not been generalized to the level of the root node in the associated DGG (which contains the default description "ANY"), the *No. of Tuples* column describes the number of tuples in the summary, and the *Index* and *Rank* columns describe

Table 6.3. Ranks assigned by $I_{Variance}$ and $I_{Simpson}$ from N-2

Summary ID	Non-ANY Attributes	No. of Tuples	$I_{Variance}$		$I_{Simpson}$	
			Index	Rank	Index	Rank
1	1	2	0.377595	1.5	0.877595	1.5
2	1	3	0.128641	5.0	0.590615	5.0
3	1	4	0.208346	3.5	0.875039	3.5
4	1	5	0.024569	10.0	0.298277	10.0
5	1	6	0.018374	12.0	0.258539	14.0
6	1	9	0.017788	13.0	0.253419	15.0
7	1	10	0.041606	8.5	0.474451	8.5
8	2	2	0.377595	1.5	0.877595	1.5
9	2	4	0.208346	3.5	0.875039	3.5
10	2	5	0.079693	6.0	0.518772	6.0
11	2	9	0.018715	11.0	0.260833	12.0
12	2	9	0.050770	7.0	0.517271	7.0
13	2	10	0.041606	8.5	0.474451	8.5
14	2	11	0.013534	14.0	0.226253	16.0
15	2	16	0.010611	17.0	0.221664	18.0
16	2	17	0.012575	15.0	0.260017	13.0
17	2	21	0.008896	18.0	0.225542	17.0
18	2	21	0.011547	16.0	0.278568	11.0
19	2	30	0.006470	19.0	0.220962	19.0
20	2	40	0.002986	20.0	0.141445	20.0
21	2	50	0.002078	21.0	0.121836	21.0
22	2	67	0.001582	22.0	0.119351	22.0

the calculated index value and the assigned rank, respectively, as determined by the corresponding measure. Table 6.3 and Tables A.1 through A.7 do not show any single-tuple summaries as these summaries are considered to contain no information and are, therefore, uninteresting by definition. The summaries in Table 6.3 and Tables A.1 through A.7 are shown in increasing order of the number of non-ANY attributes and the number of tuples in each summary, respectively.

The *Rank* column for each measure uses a ranking scheme that breaks ties in the index values by averaging the ranks and assigning the same rank to each summary involved in the tie, even though the resulting rank may be fractional. For example, if two summaries are tied when attempting to rank the fourth summary, each is given a rank of $(4 + 5)/2 = 4.5$, with the next summary ranked sixth. If instead, three summaries are tied, each is given a rank of $(4 + 5 + 6)/3 = 5.0$, with the next summary ranked seventh. The general procedure should now be clear. This ranking scheme was adopted to conform to the requirements of the Gamma correlation coefficient used to analyze the ranking similarities of the measures (described later in this section).

Table 6.3 and Tables A.1 through A.7 show there are numerous ties in the ranks assigned by each measure. For example, Summaries 1 and 8, the most interesting one- and two-attribute summaries, respectively, are tied according to each measure. This tie is an artifact of the concept hierarchies used in the discovery task (Summary 1 is shown in Table 6.4). That is, in the concept hierarchy associated with the *Province* attribute, there is a one-to-one correspondence between the concept *Canada* in Summary 8 and the concept *ANY* in Summary 1. Consequently, this results in a summary containing two non-ANY attributes being assigned the same index value as a summary containing one non-ANY attribute. All ties in Table 6.3 and Tables A.1 through A.7 result from a similar one-to-one correspondence between concepts in the concept hierarchies used, except for some of those occurring in the I_{Berger} *Rank* column of Table A.3, those being the result of the dominant tuple in many of the summaries having the same probability.

Table 6.4. Summary 1 from *N-2*

Province	DiscCode	Count
ANY	Other	8376
ANY	Computer	587

Table 6.3 and Tables A.1 through A.7 show similarities in how some of the sixteen measures rank summaries. For example, all the summaries are ranked identically by I_{Schutz}, I_{Bray}, and $I_{Whittaker}$ (in fact, it turns out that I_{Schutz}, I_{Bray}, and $I_{Whittaker}$ always rank summaries identically, and the actual value of the I_{Bray} and $I_{Whittaker}$ indexes are always identical). Also, the six most interesting summaries (i.e., summaries 1 and 8, 3 and 9, 2, and 10) are ranked identically by $I_{Variance}$, $I_{Simpson}$, $I_{Shannon}$, I_{Total}, $I_{McIntosh}$, and $I_{Kullback}$. While the four least interesting summaries (i.e., summaries 19, 20, 21, and 22) are ranked identically by $I_{Variance}$, $I_{Simpson}$, $I_{Shannon}$, I_{Total}, I_{Max}, $I_{McIntosh}$, $I_{Kullback}$, I_{Theil}, and I_{Gini}.

6.2.2. Analysis of Ranking Similarities

To quantify the extent of the ranking similarities between the sixteen measures across all seven discovery tasks, we calculate the Gamma correlation coefficient for each pair of measures. The Gamma statistic assumes that the summaries under consideration are assigned ranks according to an ordinal (i.e., rank order) scale, and is computed as the the probability that the rank ordering of two measures agree minus the probability that they disagree, divided by 1 minus the probability of ties. The value of the Gamma statistic varies in the interval $[-1, 1]$, where values near 1, 0, and -1 represent significant positive, no, and significant negative correlation, respectively.

The Gamma correlation coefficients (hereafter called the coefficients) for the seven discovery tasks are shown in Tables 6.5 and B.1 of Appendix B. In Tables 6.5 and B.1, the *Measure 1* and *Measure 2* columns describe the pairs of measures being compared, the *N-2*, *N-3*, and *N-4* columns describe the coefficients corresponding to the pairs of measures from the two-, three-, and four-attribute NSERC discovery tasks, respectively, the *C-2*, *C-3*, *C-4*, and *C-5* columns describe the coefficients corresponding to the pairs of measures from the two-, three-, four-, and five-attribute Customer discovery tasks, respectively, and the *Average* column describes the average of the coefficients for the respective group of discovery tasks. 86.43% of the coefficients in Tables 6.5 and B.1 (shown in bold) are highly significant with a *p-value* below 0.005 (68.89% of the NSERC and 99.99% of the Customer discovery tasks).

Tables 6.5 and B.1 show the ranks assigned to the summaries have a high positive correlation for some pairs of measures, as indicated by the high coefficients. For the purpose of this discussion, we consider a pair of measures to be highly correlated when the average coefficient is greater than 0.85. Thus, 35% of the pairs (i.e., 42 out of 120) are highly correlated using the 0.85 threshold. The most highly correlated pairs, with a coefficient of 1.00, are $\langle I_{Simpson}, I_{McIntosh}\rangle$, $\langle I_{Shannon}, I_{Kullback}\rangle$, $\langle I_{Schutz}, I_{Bray}\rangle$, $\langle I_{Schutz}, I_{Whittaker}\rangle$, and $\langle I_{Bray}, I_{Whittaker}\rangle$. Some of the other highly correlated pairs include $\langle I_{Total}, I_{Max}\rangle$ at 0.977215, $\langle I_{Variance}, I_{Shannon}\rangle$ and $\langle I_{Variance}, I_{Kullback}\rangle$ at 0.970057, $\langle I_{Simpson}, I_{Shannon}\rangle$ and $\langle I_{Simpson}, I_{Kullback}\rangle$ at 0.967556, $\langle I_{Shannon}, I_{McIntosh}\rangle$ and $\langle I_{McIntosh}, I_{Kullback}\rangle$ at 0.965484, $\langle I_{Variance}, I_{McIntosh}\rangle$ at 0.965473, $\langle I_{Shannon}, I_{Total}\rangle$ and $\langle I_{Total}, I_{Kullback}\rangle$ at 0.934261, $\langle I_{Variance}, I_{Total}\rangle$ at 0.933525, $\langle I_{McIntosh}, I_{Berger}\rangle$ at 0.932918, and $\langle I_{Simpson}, I_{Berger}\rangle$ at 0.932917. There are 23 other highly correlated pairs when using the 0.85 threshold.

Following careful examination of the 42 highly correlated pairs, we see that there are primarily two distinct groups of measures within which summaries are ranked similarly. One group consists of the nine measures $I_{Variance}$, $I_{Simpson}$, $I_{Shannon}$, I_{Total}, I_{Max}, $I_{McIntosh}$, I_{Berger}, $I_{Kullback}$, and I_{Gini}. The other group consists of the four measures I_{Schutz}, I_{Bray}, $I_{Whittaker}$, and $I_{MacArthur}$. There are no similarities between the two groups, as shown by the low average coefficients for all pairs consisting of one measure chosen from each group (i.e., coefficients ranging from 0.257926 to 0.438012).

Of the remaining three measures (i.e., I_{Theil}, I_{Lorenz}, and $I_{Atkinson}$), I_{Theil} is highly correlated with I_{Max}, while I_{Lorenz} and $I_{Atkinson}$ are not highly correlated with any of the other measures. There were no highly negative correlations between any of the pairs of measures.

Table 6.5. Ranking similarities for NSERC discovery tasks

Measure 1	Measure 2	Gamma Correlation Coefficient			
		N-2	*N-3*	*N-4*	*Average*
$I_{Variance}$	$I_{Simpson}$	**0.921053**	**0.939268**	**0.949392**	0.936571
$I_{Variance}$	$I_{Shannon}$	**0.956140**	**0.967554**	**0.961472**	0.961722
$I_{Variance}$	I_{Total}	**0.938597**	**0.876040**	**0.920476**	0.911704
$I_{Variance}$	I_{Max}	**0.865471**	**0.834039**	**0.898868**	0.866126
$I_{Variance}$	$I_{McIntosh}$	**0.921053**	**0.939268**	**0.949306**	0.936542
$I_{Variance}$	I_{Lorenz}	**0.692982**	**0.388519**	0.061277	0.380926
$I_{Variance}$	I_{Berger}	**0.920792**	**0.928448**	**0.858421**	0.902554
$I_{Variance}$	I_{Schutz}	0.374449	0.083368	-0.103141	0.118225
$I_{Variance}$	I_{Bray}	0.374449	0.083368	-0.103141	0.118225
$I_{Variance}$	$I_{Whittaker}$	0.374449	0.083368	-0.103141	0.118225
$I_{Variance}$	$I_{Kullback}$	**0.956140**	**0.967554**	**0.961472**	0.961722
$I_{Variance}$	$I_{MacArthur}$	**0.447368**	0.128120	-0.118625	0.152288
$I_{Variance}$	I_{Theil}	**0.736842**	**0.628120**	**0.644347**	0.669770
$I_{Variance}$	$I_{Atkinson}$	0.289474	0.009983	-0.040504	0.086318
$I_{Variance}$	I_{Gini}	**0.964912**	**0.930948**	**0.942517**	0.946126
$I_{Simpson}$	$I_{Shannon}$	**0.947368**	**0.965058**	**0.945155**	0.952527
$I_{Simpson}$	I_{Total}	**0.859649**	**0.828619**	**0.869567**	0.852612
$I_{Simpson}$	I_{Max}	**0.784753**	**0.785775**	**0.847883**	0.806137
$I_{Simpson}$	$I_{McIntosh}$	**1.000000**	**1.000000**	**1.000000**	1.000000
$I_{Simpson}$	I_{Lorenz}	**0.614035**	**0.341098**	0.016437	0.323857
$I_{Simpson}$	I_{Berger}	**0.950495**	**0.968966**	**0.878184**	0.932548
$I_{Simpson}$	I_{Schutz}	0.295154	0.035610	**-0.142141**	0.062874
$I_{Simpson}$	I_{Bray}	0.295154	0.035610	**-0.142141**	0.062874
$I_{Simpson}$	$I_{Whittaker}$	0.295154	0.035610	**-0.142141**	0.062874
$I_{Simpson}$	$I_{Kullback}$	**0.947368**	**0.965058**	**0.945155**	0.952527
$I_{Simpson}$	$I_{MacArthur}$	0.368421	0.080699	**-0.159921**	0.096400
$I_{Simpson}$	I_{Theil}	**0.657895**	**0.580699**	**0.598594**	0.612396
$I_{Simpson}$	$I_{Atkinson}$	0.210526	-0.037438	-0.080499	0.030863
$I_{Simpson}$	I_{Gini}	**0.885965**	**0.873544**	**0.893913**	0.884474
$I_{Shannon}$	I_{Total}	**0.912281**	**0.863561**	**0.909474**	0.895105
$I_{Shannon}$	I_{Max}	**0.838565**	**0.821338**	**0.886842**	0.848915
$I_{Shannon}$	$I_{McIntosh}$	**0.947368**	**0.965058**	**0.945069**	0.952498
$I_{Shannon}$	I_{Lorenz}	**0.666667**	**0.376040**	0.045485	0.362731
$I_{Shannon}$	I_{Berger}	**0.950495**	**0.946552**	**0.829860**	0.908969
$I_{Shannon}$	I_{Schutz}	0.348018	0.070800	-0.121333	0.099162
$I_{Shannon}$	I_{Bray}	0.348018	0.070800	-0.121333	0.099162
$I_{Shannon}$	$I_{Whittaker}$	0.348018	0.070800	-0.121333	0.099162
$I_{Shannon}$	$I_{Kullback}$	**1.000000**	**1.000000**	**1.000000**	1.000000
$I_{Shannon}$	$I_{MacArthur}$	0.421053	0.115641	**-0.135261**	0.133811
$I_{Shannon}$	I_{Theil}	**0.710526**	**0.615641**	**0.625154**	0.650440

Table 6.5. Ranking similarities for NSERC discovery tasks (continued)

Measure 1	Measure 2	Gamma Correlation Coefficient			
		N-2	N-3	N-4	Average
$I_{Shannon}$	$I_{Atkinson}$	0.263158	-0.002496	-0.057959	0.067568
$I_{Shannon}$	I_{Gini}	0.938596	0.908486	0.933995	0.927026
I_{Total}	I_{Max}	0.928251	0.960203	0.977615	0.955356
I_{Total}	$I_{McIntosh}$	0.859649	0.828619	0.869485	0.852584
I_{Total}	I_{Lorenz}	0.754386	0.512479	0.135663	0.467843
I_{Total}	I_{Berger}	0.871287	0.820690	0.789616	0.827198
I_{Total}	I_{Schutz}	0.436123	0.206535	-0.034610	0.202683
I_{Total}	I_{Bray}	0.436123	0.206535	-0.034610	0.202683
I_{Total}	$I_{Whittaker}$	0.436123	0.206535	-0.034610	0.202683
I_{Total}	$I_{Kullback}$	0.912281	0.863561	0.909474	0.895105
I_{Total}	$I_{MacArthur}$	0.508772	0.252080	-0.046669	0.238061
I_{Total}	I_{Theil}	0.798246	0.750416	0.711285	0.753316
I_{Total}	$I_{Atkinson}$	0.350877	0.128952	0.028824	0.169551
I_{Total}	I_{Gini}	0.973684	0.940100	0.971260	0.951681
I_{Max}	$I_{McIntosh}$	0.784753	0.785775	0.847801	0.806110
I_{Max}	I_{Lorenz}	0.865471	0.579170	0.160887	0.535176
I_{Max}	I_{Berger}	0.807107	0.780702	0.774465	0.762864
I_{Max}	I_{Schutz}	0.540541	0.268230	-0.010010	0.266254
I_{Max}	I_{Bray}	0.540541	0.268230	-0.010010	0.266254
I_{Max}	$I_{Whittaker}$	0.540541	0.268230	-0.010010	0.266254
I_{Max}	$I_{Kullback}$	0.838565	0.821338	0.886842	0.848915
I_{Max}	$I_{MacArthur}$	0.614350	0.314141	-0.022120	0.268790
I_{Max}	I_{Theil}	0.910314	0.819644	0.737552	0.822503
I_{Max}	$I_{Atkinson}$	0.408072	0.184589	0.053323	0.215328
I_{Max}	I_{Gini}	0.901345	0.900931	0.949590	0.917289
$I_{McIntosh}$	I_{Lorenz}	0.614035	0.341098	0.016392	0.324022
$I_{McIntosh}$	I_{Berger}	0.950495	0.968966	0.878189	0.932550
$I_{McIntosh}$	I_{Schutz}	0.295154	0.035610	-0.142179	0.062862
$I_{McIntosh}$	I_{Bray}	0.295154	0.035610	-0.142179	0.062862
$I_{McIntosh}$	$I_{Whittaker}$	0.295154	0.035610	-0.142179	0.062862
$I_{McIntosh}$	$I_{Kullback}$	0.947368	0.965058	0.945069	0.952498
$I_{McIntosh}$	$I_{MacArthur}$	0.368421	0.080699	-0.159958	0.096387
$I_{McIntosh}$	I_{Theil}	0.657895	0.580699	0.598523	0.612372
$I_{McIntosh}$	$I_{Atkinson}$	0.210526	-0.037438	-0.080540	0.030849
$I_{McIntosh}$	I_{Gini}	0.885965	0.873544	0.893830	0.884446
I_{Lorenz}	I_{Berger}	0.623762	0.325862	0.016844	0.322156
I_{Lorenz}	I_{Schutz}	0.647577	0.684122	0.804996	0.712232
I_{Lorenz}	I_{Bray}	0.647577	0.684122	0.804996	0.712232
I_{Lorenz}	$I_{Whittaker}$	0.647577	0.684122	0.804996	0.712232
I_{Lorenz}	$I_{Kullback}$	0.666667	0.376040	0.045485	0.362731

Table 6.5. Ranking similarities for NSERC discovery tasks (continued)

Measure 1	Measure 2	Gamma Correlation Coefficient			
		N-2	N-3	N-4	Average
I_{Lorenz}	$I_{MacArthur}$	**0.754386**	**0.732945**	**0.814898**	0.767410
I_{Lorenz}	I_{Theil}	**0.956140**	**0.747088**	**0.416304**	0.766511
I_{Lorenz}	$I_{Atkinson}$	**0.508772**	**0.601498**	**0.521697**	0.543989
I_{Lorenz}	I_{Gini}	**0.728070**	**0.454243**	0.107449	0.429921
I_{Berger}	I_{Schutz}	0.297030	0.032986	**-0.133405**	0.065537
I_{Berger}	I_{Bray}	0.297030	0.032986	**-0.133405**	0.065537
I_{Berger}	$I_{Whittaker}$	0.297030	0.032986	**-0.133405**	0.065537
I_{Berger}	$I_{Kullback}$	**0.950495**	**0.946552**	**0.829860**	0.908969
I_{Berger}	$I_{MacArthur}$	0.376238	0.070690	**-0.148850**	0.099359
I_{Berger}	I_{Theil}	**0.673267**	**0.569828**	**0.577255**	0.606783
I_{Berger}	$I_{Atkinson}$	0.188119	-0.046552	-0.067046	0.024840
I_{Berger}	I_{Gini}	**0.900990**	**0.862069**	**0.083989**	0.615683
I_{Schutz}	I_{Bray}	**1.000000**	**1.000000**	**1.000000**	1.000000
I_{Schutz}	$I_{Whittaker}$	**1.000000**	**1.000000**	**1.000000**	1.000000
I_{Schutz}	$I_{Kullback}$	0.348018	0.070800	-0.121333	0.099162
I_{Schutz}	$I_{MacArthur}$	**0.894273**	**0.919564**	**0.934562**	0.916133
I_{Schutz}	I_{Theil}	**0.638767**	**0.457897**	**0.253793**	0.450152
I_{Schutz}	$I_{Atkinson}$	**0.480176**	**0.749476**	**0.627969**	0.619207
I_{Schutz}	I_{Gini}	0.409692	0.147884	-0.063019	0.164852
I_{Bray}	$I_{Whittaker}$	**1.000000**	**1.000000**	**1.000000**	1.000000
I_{Bray}	$I_{Kullback}$	0.348018	0.070800	-0.121333	0.099162
I_{Bray}	$I_{MacArthur}$	**0.894273**	**0.919564**	**0.934562**	0.916133
I_{Bray}	I_{Theil}	**0.638767**	**0.457897**	**0.253793**	0.450152
I_{Bray}	$I_{Atkinson}$	**0.480176**	**0.749476**	**0.627969**	0.619207
I_{Bray}	I_{Gini}	0.409692	0.147884	-0.063019	0.494557
$I_{Whittaker}$	$I_{Kullback}$	0.348018	0.070800	-0.121333	0.099162
$I_{Whittaker}$	$I_{MacArthur}$	**0.894273**	**0.919564**	**0.934562**	0.916133
$I_{Whittaker}$	I_{Theil}	**0.638767**	**0.457897**	**0.253793**	0.450152
$I_{Whittaker}$	$I_{Atkinson}$	**0.480176**	**0.749476**	**0.627969**	0.619207
$I_{Whittaker}$	I_{Gini}	0.409692	0.147884	-0.063019	0.164852
$I_{Kullback}$	$I_{MacArthur}$	0.421053	0.115641	**-0.135261**	0.133811
$I_{Kullback}$	I_{Theil}	**0.710526**	**0.615641**	**0.625154**	0.650440
$I_{Kullback}$	$I_{Atkinson}$	0.263158	-0.002496	-0.057959	0.067568
$I_{Kullback}$	I_{Gini}	**0.938596**	**0.908486**	**0.933995**	0.927026
$I_{MacArthur}$	I_{Theil}	**0.710526**	**0.500000**	**0.238882**	0.483136
$I_{MacArthur}$	$I_{Atkinson}$	**0.578947**	**0.797005**	**0.642152**	0.672701
$I_{MacArthur}$	I_{Gini}	0.482456	0.193844	-0.075057	0.200414
I_{Theil}	$I_{Atkinson}$	**0.464912**	**0.370216**	**0.282533**	0.372554
I_{Theil}	I_{Gini}	**0.771930**	**0.692180**	**0.683073**	0.715728
$I_{Atkinson}$	I_{Gini}	0.324561	0.074043	0.001284	0.133296

6.2.3.   Analysis of Summary Complexity

We now discuss the complexity of the summaries ranked by the various measures. We define the *complexity* of a summary as the product of the number of tuples and the number of non-ANY attributes contained in the summary. We believe a desirable property of any ranking function is that it rank summaries with low complexity as most interesting. However, although we want to rank summaries with low complexity as most interesting, we do not want to lose the meaning or context of the data by presenting summaries that are too concise. Indeed, in previous work, domain experts agreed that more information is better than less, provided that the most interesting summaries are not too concise and remain relatively easy to understand [49].

One way to analyze the measures and evaluate whether they satisfy the guidelines of our domain experts, is to determine the complexity of those summaries considered to be of high, moderate, and low interest, as shown in Table 6.6 and Tables C.1 and C.2 of Appendix C. In Tables 6.6, C.1 and C.2, three rows (labelled *T*, *NA*, and *CI*) are used to describe the relative interestingness of summaries ranked by each measure. The first row describes the average number of tuples contained in each group of summaries, when grouped according to a three-tier scale of relative interestingness (i.e., H=High, M=Moderate, L=Low) by discovery task. The second and third rows describe the corresponding average number of non-ANY attributes in the summaries of each group and the complexity for each group, respectively. High, moderate, and low interest summaries were considered to be the top, middle,

Table 6.6.   Relative interestingness versus complexity for NSERC discovery tasks

Measure		\multicolumn{9}{c}{Relative Interestingness}								
		\multicolumn{3}{c}{N-2}	\multicolumn{3}{c}{N-3}	\multicolumn{3}{c}{N-4}
Measure		*H*	*M*	*L*	*H*	*M*	*L*	*H*	*M*	*L*
$I_{Variance}$	T	3.0	7.5	46.8	5.6	29.4	179.4	17.3	144.5	845.5
	NA	1.5	1.5	2.0	1.6	2.2	2.9	2.0	3.0	3.8
	CI	4.5	11.3	93.6	9.0	64.7	520.3	34.6	430.5	3212.9
$I_{Simpson}$	T	3.0	11.3	46.8	5.6	31.7	170.5	19.0	154.4	832.4
	NA	1.5	1.8	2.0	1.6	2.3	2.8	2.0	2.9	3.8
	CI	4.5	20.3	93.6	9.0	72.9	477.4	38.0	447.8	3163.1
$I_{Shannon}$	T	3.0	7.5	46.8	5.6	31.7	179.4	15.7	143.4	844.8
	NA	1.5	1.5	2.0	1.6	2.3	2.9	1.9	3.0	3.8
	CI	4.5	11.3	93.6	9.0	72.9	520.3	29.8	430.2	3210.2
I_{Total}	T	3.0	8.8	46.8	5.4	27.4	188.1	14.3	141.2	847.5
	NA	1.5	1.5	2.0	1.5	2.4	2.9	1.9	3.0	3.8
	CI	4.5	13.2	93.6	8.1	65.8	545.5	27.2	423.6	3220.5

Table 6.6. Relative interestingness versus complexity for NSERC discovery tasks (continued)

Measure		Relative Interestingness								
		N-2			N-3			N-4		
		H	M	L	H	M	L	H	M	L
I_{Max}	T	2.8	9.3	46.8	5.2	27.7	188.1	14.2	141.4	847.8
	NA	1.3	1.5	2.0	1.6	2.3	2.9	1.9	3.0	3.8
	CI	3.6	14.0	93.6	8.3	63.7	545.5	27.0	424.2	3221.6
$I_{McIntosh}$	T	3.0	11.3	46.8	5.6	31.7	170.5	19.0	154.4	832.4
	NA	1.5	1.8	2.0	1.6	2.3	2.8	2.0	2.9	3.8
	CI	4.5	20.3	93.6	9.0	72.9	477.4	38.0	447.8	3163.1
I_{Lorenz}	T	3.0	11.3	46.8	13.2	41.9	95.9	66.8	404.1	166.4
	NA	1.3	1.8	2.0	1.6	2.5	2.6	2.0	3.4	2.9
	CI	3.9	20.3	93.6	21.1	104.8	249.3	133.6	1373.9	482.6
I_{Berger}	T	3.0	10.5	46.8	6.4	36.1	163.4	24.4	183.7	758.7
	NA	1.5	1.5	2.0	1.5	2.4	2.8	2.0	3.2	3.7
	CI	4.5	15.8	93.6	9.6	86.6	457.5	48.8	587.8	2807.2
I_{Schutz}	T	4.0	7.3	27.0	14.6	131.4	63.8	138.0	354.9	90.8
	NA	1.0	1.8	1.8	1.6	2.8	2.3	2.1	3.5	2.5
	CI	4.0	13.1	48.6	23.4	367.9	146.7	289.8	1242.2	227.0
I_{Bray}	T	4.0	7.3	27.0	14.6	131.4	63.8	138.0	354.9	90.8
	NA	1.0	1.8	1.8	1.6	2.8	2.3	2.1	3.5	2.5
	CI	4.0	13.1	48.6	23.4	367.9	146.7	289.8	1242.2	227.0
$I_{Whittaker}$	T	4.0	7.3	27.0	14.6	131.4	63.8	138.0	354.9	90.8
	NA	1.0	1.8	1.8	1.6	2.8	2.3	2.1	3.5	2.5
	CI	4.0	13.1	48.6	23.4	367.9	146.7	289.8	1242.2	227.0
$I_{Kullback}$	T	3.0	7.5	46.8	5.6	31.7	179.4	15.7	143.4	844.8
	NA	1.5	1.5	2.0	1.6	2.3	2.9	1.9	3.0	3.8
	CI	4.5	11.3	93.6	9.0	72.9	520.3	29.8	430.2	3210.2
$I_{MacArthur}$	T	3.8	7.3	42.0	14.5	96.7	88.3	118.8	345.8	89.7
	NA	1.3	1.8	2.0	1.6	2.6	2.5	2.1	3.5	2.6
	CI	4.9	13.1	84.0	23.2	251.4	220.8	249.5	1210.3	233.2
I_{Theil}	T	3.0	9.5	46.8	5.7	27.6	183.9	17.8	186.3	721.2
	NA	1.3	1.8	2.0	1.6	2.4	2.9	1.9	3.0	3.7
	CI	3.9	17.1	93.6	9.1	66.2	533.3	33.8	558.9	2668.4
$I_{Atkinson}$	T	5.3	10.0	27.3	16.6	117.6	45.1	221.3	185.2	519.7
	NA	1.5	1.8	1.8	1.9	2.3	2.3	2.4	3.0	3.1
	CI	8.0	18.0	49.1	31.5	270.5	103.7	531.1	555.6	1611.1
I_{Gini}	T	3.0	8.8	46.8	5.6	27.5	185.4	14.7	141.7	847.5
	NA	1.5	1.5	2.0	1.6	2.2	2.9	1.9	3.0	3.8
	CI	4.5	13.2	93.6	9.0	60.5	537.7	27.9	425.1	3220.5

and bottom 20%, respectively, of summaries as ranked by each measure. The *N-2*, *N-3*, and *N-4* discovery tasks generated sets containing 22, 70, and 214 summaries, respectively, while the *C-2*, *C-3*, *C-4*, and *C-5* discovery tasks generated sets containing 43, 91, 155, and 493 summaries, respectively. Thus, the complexity of the summaries from the *N-2*, *N-3*, and *N-4* discovery tasks is based upon four, 14, and 43 summaries, respectively, while the complexity of the summaries from the *C-2*, *C-3*, *C-4*, and *C-5* discovery tasks is based upon nine, 18, 31, and 97 summaries, respectively.

Tables 6.6, C.1, and C.2 show that in most cases the complexity is lowest for the most interesting summaries and highest for the least interesting summaries. For example, the complexity of summaries determined to be of high, moderate, and low interest, when ranked by $I_{Variance}$, are 4.5, 11.3, and 93.6 from *N-2*, 9.0, 64.7, and 520.3 from *N-3*, and 34.6, 430.5, and 3212.9 from *N-4*. The only exceptions are the summaries from *N-3* and *N-4* ranked by I_{Lorenz}, I_{Schutz}, I_{Bray}, $I_{Whittaker}$, $I_{MacArthur}$, and $I_{Atkinson}$. For example, the complexity of summaries determined to be of high, moderate, and low interest, when ranked by I_{Schutz}, are 23.4, 367.9, and 146.7 from *N-3* and 289.8, 1242.2, and 227.0 from *N-4*. That is, the summaries considered to be of moderate interest have the highest complexity. Also, from *N-4*, the summary considered to be of least interest has the lowest complexity. There were no exceptions from the *C-2*, *C-3*, *C-4*, and *C-5* discovery tasks.

A graphical comparison of the complexity of the summaries ranked by the sixteen measures is shown in the graphs of Figures 6.5 through 6.8. In Figures 6.5 through 6.8, the horizontal and vertical axes describe the measures and the complexity, respectively. In Figures 6.5 and 6.6, the horizontal rows of bars labelled *High*, *Moderate*, and *Low* correspond to the top, middle, and bottom 20%, respectively, of summaries ranked by each measure. Figures 6.5 and 6.6 provide a graphical representation of the results described in the previous paragraph.

In Figures 6.7 and 6.8, each horizontal row of bars corresponds to the complexity of the most interesting summaries (i.e., top 20%) from a particular discovery task. The backmost horizontal row of bars corresponds to the average complexity for a particular measure. Both figures show a maximum complexity on the vertical axes of 60.0, although the complexity of the most interesting summaries ranked by I_{Lorenz}, I_{Schutz}, I_{Bray}, $I_{Whittaker}$, $I_{MacArthur}$, and $I_{Atkinson}$ in *N-4* exceed this value (i.e., 133.6, 289.8, 289.8, 289.8, 249.5, and 531.1, respectively). When the measures are ordered by complexity, from lowest to highest, they are ordered according to Figure 6.7, as follows (position in the ordering is shown in parentheses): I_{Max} (1), I_{Total} (2), I_{Gini} (3), $I_{Shannon}$ and $I_{Kullback}$ (4), I_{Theil} (5), $I_{Variance}$ (6), $I_{Simpson}$ and $I_{McIntosh}$ (7), I_{Berger} (8), I_{Lorenz} (9), $I_{MacArthur}$ (10), I_{Schutz}, I_{Bray}, and $I_{Whittaker}$ (11), and $I_{Atkinson}$ (12). They are ordered according to Figure 6.8, as follows: I_{Total} (1), I_{Max} (2), I_{Berger} (3), $I_{Variance}$, $I_{Simpson}$, $I_{Shannon}$, $I_{McIntosh}$, and I_{Gini} (4), $I_{Kullback}$ (5), I_{Lorenz} (6), $I_{MacArthur}$ (7), $I_{Atkinson}$ (8), I_{Schutz}, I_{Bray}, and $I_{Whittaker}$ (9).

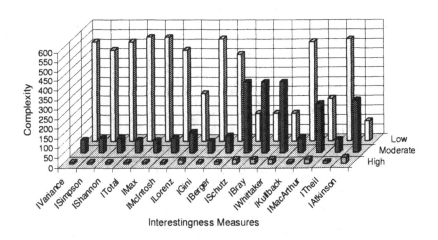

Figure 6.5. Relative complexity of summaries within *N-3*

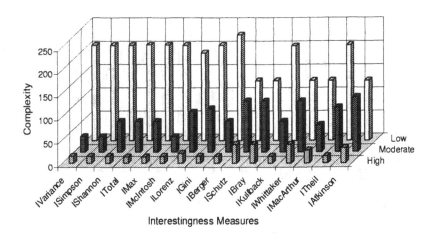

Figure 6.6. Relative complexity of summaries within *C-4*

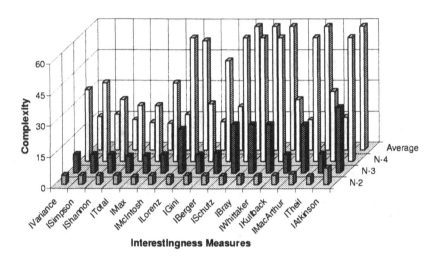

Figure 6.7. Relative complexity of summaries between NSERC discovery tasks

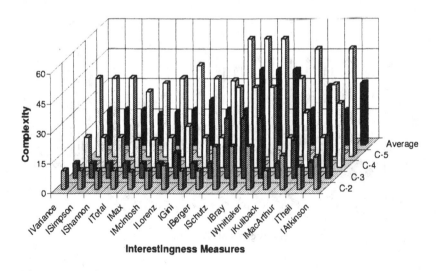

Figure 6.8. Relative complexity of summaries between Customer discovery tasks

6.2.4. Distribution of Index Values

We now analyze the distribution of the index values generated by each of the sixteen measures. We test the series of vectors shown in Table 6.7, where index values for 16,928 vectors (i.e., all possible ordered arrangements of a population of 50 objects among 10 classes) and 2,611 vectors (i.e., all possible ordered arrangements of a population of 50 objects among 5 classes) were generated. The choice of vectors to evaluate here was made somewhat arbitrarily, but it does provide a large, controlled population of index values in which a gradual change in evenness occurs from the most highly skewed distribution in the first vector, to the uniform distribution in the last vector.

Table 6.7. Ordered arrangements of two populations

50 objects / 10 classes	*50 objects / 5 classes*
(41, 1, 1, 1, 1, 1, 1, 1, 1, 1)	(46, 1, 1, 1, 1)
(40, 2, 1, 1, 1, 1, 1, 1, 1, 1)	(45, 2, 1, 1, 1)
(39, 3, 1, 1, 1, 1, 1, 1, 1, 1)	(44, 3, 1, 1, 1)
⋮	⋮
(6, 6, 5, 5, 5, 5, 5, 5, 4, 4)	(11, 11, 10, 10, 8)
(6, 5, 5, 5, 5, 5, 5, 5, 5, 4)	(11, 10, 10, 10, 9)
(5, 5, 5, 5, 5, 5, 5, 5, 5, 5)	(10, 10, 10, 10, 10)

Histograms of the absolute frequencies of the index values were generated for each measure for the sequences in Table 6.7. However, due to space limitations, we are able to show only two histograms, those generated by $I_{Variance}$ and I_{Schutz}, shown in Figures 6.9 and 6.10, respectively, and corresponding to the population of 50 objects among 10 classes. In Figures 6.9 and 6.10, the horizontal and vertical axes of each graph describe intervals of the index values generated by a measure and the number of index values that fall in each interval, respectively. For example, the histogram for $I_{Variance}$ in in Figure 6.9, shows that 68 index values were generated on the interval $(0.000, 0.0009]$, 1,106 on $(0.0009, 0.003]$, 2,464 on $(0.003, 0.005]$, 3,006 on $(0.005, 0.007]$, 2,581 on $(0.007, 0.008]$, 2,055 on $(0.008, 0.010]$, 1,549 on $(0.010, 0.012]$, and 4,099 on the remaining intervals in $(0.012, 0.065]$. A curve describing the standard normal distribution (SND) of the index values is superimposed over the observed frequencies.

To provide a summary description of all the histograms generated, we can use the skewness and kurtosis for the distribution of index values. *Skewness* is a measure of the symmetry of a distribution. It has a value of zero when the distribution is a symmetrical curve (i.e., as in a SND). If the skewness is

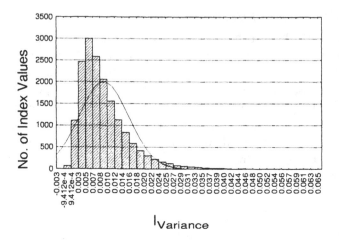

Figure 6.9. Histogram of index value frequencies for $I_{Variance}$

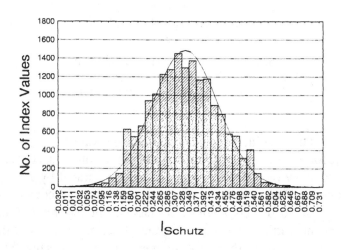

Figure 6.10. Histogram of index value frequencies for I_{Schutz}

different from zero, then the distribution is asymmetrical. A positive (negative) value indicates the index values are clustered more to the left (right) of the mean, with most of the extreme index values to the right (left) of the mean. In general, for positive (negative) skewness, we have mode $\leq$ median $\leq$ mean (mean $\leq$ median $\leq$ mode). *Kurtosis* is a measure of the relative peakedness of a distribution and indicates the extent to which outliers cause the distribution to differ from the SND. When a distribution follows the SND, it has value of zero. When the value is greater than (less than) zero, the distribution has a sharper (flatter) peak than the SND and is more (less) prone to containing outliers.

The skewness and kurtosis for all measures are shown in Table 6.8. In Table 6.8, mnemonics are provided as an aid to interpreting the curve described by the actual skewness and kurtosis values. The skewness mnemonics describe the symmetry of the frequency distribution in relation to the mean (i.e., AL = asymmetrical left, AR = asymmetrical right, NS = near symmetrical, and S = symmetrical) and the kurtosis mnemonics describe the relative peakedness of the frequency distribution in relation to the SND (i.e., SP = sharp peaked, NSN = near standard normal, MP = more peaked, and LP = less peaked). For example, the histogram for $I_{Variance}$ in Figure 6.9 has a skewness and kurtosis of approximately 1.8 and 5.6, respectively. This means that the distribution of index values is asymmetrical to the left of the mean (i.e., AL) and more sharply peaked than the SND (i.e., SP). Similarly, the histogram for I_{Schutz} in Figure 6.10 has a distribution of index values that is near symmetrical (i.e., NS) and less peaked than the SND (i.e., LP). The other measures in Table 6.8 can also be interpreted similarly. The I_{Max} measure generates the same index value for each distribution in both populations, so it has no variance and a histogram was not generated.

Table 6.8. Skewness and kurtosis of the index values for the two populations

Measure	50 objects / 10 classes				50 objects / 5 classes			
	Skewness		Kurtosis		Skewness		Kurtosis	
$I_{Variance}$	1.84421	AL	5.571732	SP	1.55959	AL	3.273237	SP
$I_{Simpson}$	1.84421	AL	5.571732	SP	1.55959	AL	3.273237	SP
$I_{Shannon}$	-0.95761	AR	1.357844	MP	-1.03452	AR	1.391038	MP
I_{Total}	-0.95761	AR	1.357844	MP	-1.03452	AR	1.391038	MP
I_{Max}	-	-	-	-	-	-	-	-
$I_{McIntosh}$	-1.24351	AR	2.317341	SP	-1.13072	AR	1.496420	SP
I_{Lorenz}	0.14435	S	-0.232495	NSN	0.02128	S	-0.317871	NSN
I_{Gini}	-0.14435	S	-0.232495	NSN	-0.02128	S	-0.317871	NSN
I_{Berger}	0.97607	AL	1.139526	SP	0.75039	AL	0.264196	SP
I_{Schutz}	0.13192	NS	-0.130277	LP	0.27521	NS	-0.076436	LP
I_{Bray}	-0.13192	NS	-0.130277	LP	-0.27521	NS	-0.076436	LP
$I_{Whittaker}$	-0.13192	NS	-0.130277	LP	-0.27521	NS	-0.076436	LP
$I_{Kullback}$	-0.95761	AR	1.357884	MP	-1.03452	AR	1.391038	MP
$I_{MacArthur}$	0.68369	AL	0.485805	MP	0.86586	AL	0.883313	MP
I_{Theil}	-0.05563	S	-0.236451	NSN	0.68371	AL	1.112360	MP
$I_{Atkinson}$	0.16650	NS	-0.422023	LP	0.30949	AL	-0.476633	LP

We now determine the number of index values generated by each measure that are less than or greater than the middle index value (i.e., $(minimum + maximum)/2$), and less than or greater than the median (i.e., the value for

Table 6.9. Distribution of index values for 50 objects among 10 classes

Measure	Minimum	Maximum	Middle	< Middle	> Middle	Median
$I_{Variance}$	0.0	0.064	0.032	16761	167	0.007911
$I_{Simpson}$	0.1	0.676	0.388	16761	167	0.1712
$I_{Shannon}$	1.250664	3.321928	2.286295	613	16315	2.860161
I_{Total}	12.506635	33.219281	22.86296	613	16315	28.601607
I_{Max}	-	-	-	-	-	-
$I_{McIntosh}$	0.207096	0.7964	0.50175	509	16419	0.682799
I_{Lorenz}	0.214	0.55	0.37	12353	4575	0.338
I_{Gini}	0.107	0.275	0.185	4786	12142	0.169
I_{Berger}	0.14	0.82	0.46	15836	1092	0.28
I_{Schutz}	0.0	0.72	0.36	10751	6177	0.34
I_{Bray}	0.28	1.0	0.64	7549	9379	0.66
$I_{Whittaker}$	0.28	1.0	0.64	7549	9379	0.66
$I_{Kullback}$	1.250664	3.321928	2.286295	613	16315	2.860161
$I_{MacArthur}$	0.0	0.420842	0.21042	15683	1245	0.114606
I_{Theil}	0.0	2.141432	1.07072	5550	11378	1.21593
$I_{Atkinson}$	0.0	0.71	0.35503	11432	5496	0.296977

Table 6.10. Distribution of index values for 50 objects among 5 classes

Measure	Minimum	Maximum	Middle	< Middle	> Middle	Median
$I_{Variance}$	0.0	0.162	0.081	2507	104	0.0258
$I_{Simpson}$	0.2	0.848	0.524	2507	104	0.3032
$I_{Shannon}$	0.562179	2.321928	1.44205	164	2447	1.940238
I_{Total}	2.810896	11.609640	7.210268	164	2447	9.701192
I_{Max}	-	-	-	-	-	-
$I_{McIntosh}$	0.092165	0.643839	0.36800	200	2411	0.523381
I_{Lorenz}	0.24	0.6	0.42	1496	1115	0.412
I_{Gini}	0.12	0.300	0.21	1183	1428	0.0.206
I_{Berger}	0.2	0.92	0.56	2180	431	0.42
I_{Schutz}	0.0	0.72	0.36	1850	761	0.3
I_{Bray}	0.28	1.0	0.64	939	1672	0.7
$I_{Whittaker}$	0.28	1.0	0.64	939	1672	0.7
$I_{Kullback}$	0.562179	2.321928	1.44205	164	2447	1.940238
$I_{MacArthur}$	0.0	0.427524	0.213765	2425	186	0.099571
I_{Theil}	0.0	1.759749	0.879875	2357	254	0.566115
$I_{Atkinson}$	0.0	0.784944	0.39247	1964	647	0.283374

which 50% of the generated index values lie below and 50% lie above). Our belief is that a useful measure of interestingness should generate index values that are reasonably distributed throughout the range of possible values (such as in a SND or uniform distribution). Again, we analyze the index values generated from the two populations shown in Table 6.7, with the results shown in Tables 6.9 and 6.10. In Tables 6.9 and 6.10, the *Minimum* and *Maximum* columns describe the minimum and maximum index values generated by each measure, respectively, the *Middle* column describes the middle index value, the < *Middle* and > *Middle* columns describe the number of index values less than or greater than the middle index value, respectively, and the *Median* column describes the median index value. For example, for the $I_{Variance}$ measure, the minimum and maximum index values are 0.0 and 0.064, respectively, the middle index value is 0.032, 16,761 (167) index values lie below (above) the middle index value, and the median index value is 0.00791. The distribution of index values in Tables 6.9 and 6.10 is highly skewed about the middle and median values for most of the measures. Isolated exceptions include I_{Bray} and $I_{Whittaker}$ in Table 6.9, and I_{Lorenz} and I_{Gini} in Table 6.10.

We now identify those vectors whose index values are at or near the middle index value. The results are shown in Table 6.11. Where two vectors are shown for

Table 6.11. Vectors at the middle index value for two populations

Measure	50 objects / 10 classes	50 objects / 5 classes
$I_{Variance}$	(30, 6, 4, 3, 2, 1, 1, 1, 1, 1)	(35, 8, 4, 2, 1)
$I_{Simpson}$	(30, 6, 4, 3, 2, 1, 1, 1, 1, 1)	(35, 8, 4, 2, 1)
$I_{Shannon}$	(22, 13, 7, 2, 1, 1, 1, 1, 1, 1)	(23, 23, 2, 1, 1)
I_{Total}	(22, 13, 7, 2, 1, 1, 1, 1, 1, 1)	(23, 23, 2, 1, 1)
I_{Max}	-	-
$I_{McIntosh}$	(26, 10, 4, 3, 2, 1, 1, 1, 1, 1)	(33, 6, 5, 3, 3)
	(27, 7, 2, 2, 2, 2, 2, 2, 2, 2)	(29, 18, 1, 1, 1)
I_{Lorenz}	(9, 8, 8, 8, 8, 5, 1, 1, 1, 1)	(26, 8, 6, 5, 5)
	(23, 3, 3, 3, 3, 3, 3, 3, 3, 3)	(16, 16, 16, 1, 1)
I_{Gini}	(9, 8, 8, 8, 8, 5, 1, 1, 1, 1)	(26, 8, 6, 5, 5)
	(23, 3, 3, 3, 3, 3, 3, 3, 3, 3)	(16, 16, 16, 1, 1)
I_{Berger}	(23, 19, 1, 1, 1, 1, 1, 1, 1, 1)	(28, 9, 9, 3, 1)
	(23, 3, 3, 3, 3, 3, 3, 3, 3, 3)	(16, 16, 16, 1, 1)
I_{Schutz}	(23, 5, 5, 5, 5, 3, 1, 1, 1, 1)	(28, 9, 9, 3, 1)
	(23, 3, 3, 3, 3, 3, 3, 3, 3, 3)	(16, 16, 16, 1, 1)
I_{Bray}	(23, 5, 5, 5, 5, 3, 1, 1, 1, 1)	(28, 9, 9, 3, 1)
	(23, 3, 3, 3, 3, 3, 3, 3, 3, 3)	(16, 16, 16, 1, 1)
$I_{Whittaker}$	(23, 5, 5, 5, 5, 3, 1, 1, 1, 1)	(28, 9, 9, 3, 1)
	(23, 3, 3, 3, 3, 3, 3, 3, 3, 3)	(16, 16, 16, 1, 1)
$I_{Kullback}$	(22, 13, 7, 2, 1, 1, 1, 1, 1, 1)	(23, 23, 2, 1, 1)
$I_{MacArthur}$	(27, 6, 5, 2, 2, 2, 2, 2, 1, 1)	(35, 7, 4, 2, 2)
I_{Theil}	(21, 5, 5, 5, 5, 5, 1, 1, 1, 1)	(25, 18, 3, 2, 2)
$I_{Atkinson}$	(13, 13, 6, 6, 5, 2, 2, 1, 1, 1)	(23, 18, 5, 2, 2)

one measure, meaning there were multiple vectors assigned the same index value, the first (second) vector is the least (most) skewed vector at or near the middle index value. Most of the vectors in the *50 objects among 10 classes* column seem too skewed (on an intuitive level, at least) to be assigned the middle index value. Possible exceptions may be the least skewed vectors for I_{Lorenz} and I_{Gini}. The histograms for these measures are symmetrical about the mean, meaning the middle value is not much different from the mean, so this seems reasonable. Another possible exception is the vector for $I_{Atkinson}$. All the vectors in the *50 objects among 5 classes* column seem too skewed to be assigned the middle index value.

Chapter 7

CONCLUSION

The objective of this book was to develop and evaluate a technique for ranking the interestingness of discovered patterns in data. Four goals were realized in obtaining this objective:

- DGGs, a data structure for describing and guiding the generation of summaries from databases, were introduced.

- Serial and parallel algorithms for traversing the generalization space described by DGGs were introduced and evaluated.

- The use of diversity measures as measures of interestingness for summaries generated from databases was introduced and evaluated.

- A preliminary foundation for a theory of interestingness within the context of ranking the interestingness of summaries generated from databases was developed.

7.1. Summary

In Chapter 1, we introduced the problem of mining summaries from databases using DGGs. We showed how a DGG can be associated with a single attribute to describe and guide the generation of all possible summaries consistent with the DGG, and how summaries can be generated when multiple attributes are associated with DGGs. We also introduced the problem of ranking the interestingness of the summaries generated using diversity measures as heuristic measures of interestingness.

In Chapter 2, we presented a survey of classical data mining techniques and algorithms, and briefly described some of their important characteristics. Classical data mining techniques that continue to be widely used and researched include classification rules, association rules, clusters, sequential patterns, time series, contingency tables, and generalized relations. A detailed survey of relevant interestingness measures was also presented. These include probabilistic, syntactic, and distance measures for determining the interestingness of both single rules and complete rule sets. Both objective and subjective measures are described.

In Chapter 3, we defined DGGs and summaries, and described serial and parallel versions of the All_Gen algorithm. The operation of the *All_Gen* algorithm was thoroughly described via detailed walkthroughs of both versions. Generally, the *All_Gen* algorithm partitions discovery tasks across multiple processors by assigning a unique combination of paths, consisting of one path through the DGG associated with each attribute, to each processor.

In Chapter 4, we described sixteen well-known diversity measures, drawn from various disciplines, that we considered for use as heuristic measures of interestingness. Diversity measures were considered for this purpose because they have been successfully and frequently applied in several areas of the physical, social, ecological, management, information, and computer sciences. The selected diversity measures share three important properties that makes them simple and fast to compute. A detailed example of ranking a sample summary was presented for each measure.

In Chapter 5, we proposed five principles of interestingness for diversity measures used as heuristic measures of interestingness for ranking summaries. The theoretical results showed that six measures satisfy all of the proposed principles and appear to hold some promise as measures for ranking the interestingness of summaries generated from databases. These six measures include $I_{Variance}$, $I_{Simpson}$, $I_{Shannon}$, I_{Total}, $I_{McIntosh}$, and $I_{Kullback}$. Within this group of measures, two are from statistics (i.e., $I_{Variance}$ and $I_{Simpson}$), one is from ecology (i.e., $I_{McIntosh}$), and three are from information theory ($I_{Shannon}$, I_{Total}, and $I_{Kullback}$). The remaining ten measures did not perform as well, failing to satisfy one or more of the proposed principles.

In Chapter 6, the experimental results showed that the parallel All_Gen algorithm is an effective technique for efficient generation of all possible summaries consistent with a set of DGGs. It was shown that as the number of nodes or paths is increased in a DGG, or the number of attributes in a discovery task is increased, the traversal time of the generalization space also increases. However, it was also shown that increasing the number of processors assigned to a discovery task can significantly reduce the time required to obtain results.

Experimental results also showed that for selected databases, the order in which some of the measures rank summaries is highly correlated. It was shown that in most cases for the sixteen measures evaluated, the less complex summaries (i.e., those with fewer cells) are ranked as most interesting and the more complex summaries are ranked as least interesting. Eight measures showed the least complexity for the most interesting summaries. These include the I_{Max} and I_{Theil} measures, which do not satisfy the proposed principles. However, the six remaining measures were those that satisfied all of the proposed principles, namely $I_{Variance}, I_{Simpson}, I_{Shannon}, I_{Total}, I_{McIntosh},$ and $I_{Kullback}.$

The experimental results show that the distribution of index values, in relation to the mean, is least skewed for $I_{Lorenz}, I_{Gini}, I_{Schutz}, I_{Bray},$ and $I_{Whittaker}.$ The remaining eight measures were skewed in relation to the mean, and more or less peaked than the SND. The distribution of the index values is also highly skewed, in relation to the middle and median values, for most of the measures. Possible exceptions include $I_{Lorenz}, I_{Gini}, I_{Bray},$ and $I_{Whittaker}.$

Finally, given the theoretical and experimental results, we have six principled, candidate diversity measures for ranking summaries from which to choose (i.e., $I_{Variance}, I_{Simpson}, I_{Shannon}, I_{Total}, I_{McIntosh},$ and $I_{Kullback}$). Since each measure represents an alternative definition for diversity, the choice of which measure to use may make a difference. To get a quick indication of how the six measures may rank summaries, we suggest using the I_{Berger} measure. Although I_{Berger} only matches four of the proposed principles, it does have a relatively high coefficient of ranking similarity with the six principled measures. We suggest the use of I_{Berger} because it is calculated simply as the proportional dominance of the tuple in a summary with the highest probability of occurrence. This can be calculated quickly with no passes through the summaries being ranked. The rank order can then be refined by applying the principled measure determined to be most suitable for the application. Unfortunately, determining the most suitable measure will be a matter of trial-and-error. Similarly, applying new diversity measures (i.e., a measure not evaluated in our study) may also be a matter of trial-and-error. Thus, when choosing any candidate interestingness measure for ranking summaries, determine which of the proposed principles is satisfied, and then using this knowledge, judge the suitability of the candidate interestingness measure for the intended application.

7.2. Areas for Future Research

Considerable research remains to be done in the application of diversity measures to the problem of ranking the interestingness of summaries generated from databases. We see six major areas for future research. First, other diversity measures need to be evaluated to determine their suitability for ranking the interestingness of summaries generated from databases. There is certainly no shortage of possible candidates in the literature. Possible heuristics include *Pratt's Generalized Measure* [111], *Ray and Singer's CON-Index* [117], *Hurlbert's PIE* [76], *Gaston's Measure* [40], *Allison's Modified Squared Variation Coefficient* [10], the *Yule Characteristic* [31], the *Variance of the Logarithm* [31], and *Hill's Index* [74].

Second, Bayesian approaches should be evaluated. In the work presented here, the uniform distribution q was used in the calculation of the I_{Bray}, $I_{Whittaker}$, $I_{MacArthur}$, and $I_{Kullback}$ measures. Actually, any historical or expected distribution could have been used instead of q. For example, $I_{Kullback}$ could be used to measure the distance of the actual distribution from that of an unknown distribution. Since we have no information regarding the unknown distribution, we assume it will have some distribution s. We then use a Bayesian approach to modify our beliefs regarding the unknown distribution s, known as the *prior distribution*, before measuring the KL-distance. We modify our beliefs based upon accumulated evidence, such as the actual distribution p. On the basis of p, we modify the prior distribution to obtain a new one called the *posterior distribution*. *Bayes Theorem* tells us what to do to obtain the posterior distribution, and is given by:

$$P(A_i|X) \;=\; \frac{P(A_i|H)P(X|A_i)}{\sum_{j=1}^{n} P(A_j|H)P(X|A_j)}$$

where the A_i's are a set of mutually exclusive and exhaustive alternatives, $P(A_i|H)$ is the prior probability of A_i, X is the actual summary with distribution p, and $P(X|A_i)$ is the probability associated with A_i in the actual summary X. $P(A_i|X)$ is the *posterior probability* of A_i. Now the posterior probabilities for s are given by $P(t_1|s)$, $P(t_2|s)$, ..., $P(t_m|s)$. So, from the posterior probabilities of the t_i's in s, we have $q(t_1) = P(t_1|s)$, $q(t_2) = P(t_2|s)$, ..., $q(t_m) = P(t_m|s)$. If we now let p be the prior distribution, we can determine the value of $I_{Kullback}$ using p and the posterior distribution q.

Third, a framework will be developed for pruning the number of summaries under consideration. The framework will utilize the chi-square test of independence to identify those summaries containing a statistically significant association between attributes. Applications involving two-dimensional contingency tables abound [43, 87]. The framework should include the capability to analyze multi-dimensional contingency tables, possibly containing up to four or five attributes, where both partial and multiple associations are considered. After pruning, the remaining summaries will be ranked according to the heuristic measures of interestingness.

Fourth, principles of interestingness for comparing summaries generated from different databases need to be developed and evaluated. The following four principles, stated informally, may have some applicability. First, the *Universal Increase Principle* states that the interestingness of a summary decreases when the count in each tuple is increased by a positive constant. Second, the *Population Sensitivity Principle* states that an interestingness measure should consider the absolute differences within populations, not the proportional differences. That is, when all the tuple counts in a summary are increased by the same proportion, then the interestingness should increase. Third, the *Dom-*

inance Principle states that when the count of one tuple in a summary is increased, then interestingness increases. And finally, the *Subordinance Principle*, states that when the count of a tuple in a summary is decreased, then interestingness increases.

Fifth, the generation of duplicate summaries should be eliminated or minimized. For example, whenever available, domain constraints or a statistical evaluation of paths could prevent considering unwanted conjunctions of attribute values, resulting in a net increase in speed of the *All_Gen* algorithm and a net decrease in the number of summaries that need to be considered.

Finally, it would be interesting to analyze the similarities of highly ranked summaries within, and between, different measures, to identify common characteristics shared by the summaries (i.e., in terms of the levels of generalization of the common attributes).

References

[1] R. Agrawal, J. Gehrke, D. Gunopulos, and P. Raghavan. Automatic subspace clustering of high dimensional data for data mining applications. In *Proceedings of the ACM SIGMOD International Conference on Management of Data (SIGMOD'98)*, pages 94–105, June 1998.

[2] R. Agrawal, T. Imielinski, and A. Swami. Mining association rules between sets of items in large databases. In *Proceedings of the ACM SIGMOD International Conference on the Management of Data (SIGMOD'93)*, pages 207–216, Washington, D.C., May 1993.

[3] R. Agrawal, K. Lin, H.S. Sawhney, and K. Shim. Fast similarity search in the presence of noise, scaling, and translation in time-series databases. In *Proceedings of the 21th International Conference on Very Large Databases (VLDB'95)*, pages 490–501, Zurich, Switzerland, September 1995.

[4] R. Agrawal, H. Mannila, R.Srikant, H.Toivonen, and A.I. Verkamo. Fast discovery of association rules. In U.M. Fayyad, G. Piatetsky-Shapiro, P. Smyth, and R. Uthurusamy, editors, *Advances in Knowledge Discovery and Data Mining*, pages 307–328, Menlo Park, California, 1996. AAAI Press/MIT Press.

[5] R. Agrawal and J.C. Shafer. Parallel mining of association rules. *IEEE Transactions on Knowledge and Data Engineering*, 8(6):962–969, December 1996.

[6] R. Agrawal and R. Srikant. Fast algorithms for mining association rules. In *Proceedings of the 20th International Conference on Very Large Databases (VLDB'94)*, pages 487–499, Santiago, Chile, September 1994.

[7] R. Agrawal and R. Srikant. Mining sequential patterns. In *Proceedings of the 11th International Conference on Data Engineering*, pages 3–14, Taipei, Taiwan, March 1995.

[8] R.V. Alatalo. Problems in the measurement of evenness in ecology. *Oikos*, 37(2):199–204, 1981.

[9] P.D. Allison. Measures of inequality. *American Sociological Review*, 43:865–880, 1978.

[10] P.D. Allison. Inequality and scientific productivity. *Social Studies of Science*, 10:163–179, 1980.

[11] K. Alsabti, S. Ranka, and V. Singh. CLOUDS: A decision tree classifier for large datasets. In *Proceedings of the Fourth International Conference on Knowledge Discovery and Data Mining (KDD'98)*, pages 2–8, New York, New York, August 1998.

[12] A.B. Atkinson. On the measurement of inequality. *Journal of Economic Theory*, 2:244–263, 1970.

[13] M. Attaran and M. Zwick. An information theory approach to measuring industrial diversification. *Journal of Economic Studies*, 16:19–30, 1989.

[14] W.H. Berger and F.L. Parker. Diversity of planktonic forminifera in deep-sea sediments. *Science*, 168:1345–1347, 1970.

[15] I. Bournaud and J.-G. Ganascia. Accounting for domain knowledge in the construction of a generalization space. In *Proceedings of the Third International Conference on Conceptual Structures*, pages 446–459. Springer-Verlag, August 1997.

[16] J.R. Bray and J.T. Curtis. An ordination of the upland forest communities of southern Wisconsin. *Ecological Monographs*, 27:325–349, 1957.

[17] S. Brin, R. Motwani, and C. Silverstein. Beyond market baskets: Generalizing association rules to correlations. In *Proceedings of the ACM SIGMOD International Conference on Management of Data (SIGMOD'97)*, pages 265–276, May 1997.

[18] S. Brin, R. Motwani, J.D. Ullman, and S. Tsur. Dynamic itemset counting and implication rules for market basket data. In *Proceedings of the ACM SIGMOD International Conference on Management of Data (SIGMOD'97)*, pages 255–264, May 1997.

[19] O. Buchter and R. Wirth. Discovery of association rules over ordinal data: A new and faster algorithm and its application to basket analysis. In X. Wu, R. Kotagiri, and K. Korb, editors, *Proceedings of the Second Pacific-Asia Conference on Knowledge Discovery and Data Mining (PAKDD'98)*, pages 36–47, Melbourne, Australia, April 1998.

[20] L. Bulla. An index of evenness and its associated diversity measure. *Oikos*, 70(1):167–171, 1994.

[21] Y. Cai, N. Cercone, and J. Han. Attribute-oriented induction in relational databases. In G. Piatetsky-Shapiro and W. Frawley, editors, *Knowledge Discovery in Databases*, pages 213–228, Cambridge, Massachusetts, 1991. AAAI/MIT Press.

[22] C.L. Carter and H.J. Hamilton. Fast, incremental generalization and regeneralization for knowledge discovery from databases. In *Proceedings of the 8th Florida Artificial Intelligence Symposium*, pages 319–323, Melbourne, Florida, April 1995.

[23] C.L. Carter and H.J. Hamilton. A fast, on-line generalization algorithm for knowledge discovery. *Applied Mathematics Letters*, 8(2):5–11, 1995.

[24] C.L. Carter and H.J. Hamilton. Performance evaluation of attribute-oriented algorithms for knowledge discovery from databases. In *Proceedings of the Seventh IEEE International Conference on Tools with Artificial Intelligence (ICTAI'95)*, pages 486–489, Washington, D.C., November 1995.

[25] C.L. Carter and H.J. Hamilton. Efficient attribute-oriented algorithms for knowledge discovery from large databases. *IEEE Transactions on Knowledge and Data Engineering*, 10(2):193–208, March/April 1998.

[26] C.L. Carter, H.J. Hamilton, and N. Cercone. Share-based measures for itemsets. In J. Komorowski and J. Zytkow, editors, *Proceedings of the First European Conference on the Principles of Data Mining and Knowledge Discovery (PKDD'97)*, pages 14–24, Trondheim, Norway, June 1997.

[27] P. Clark and T. Niblett. The CN2 induction algorithm. *Machine Learning*, 3:261–283, 1989.

[28] H. Dalton. The measurement of the inequality of incomes. *Economic Journal*, 30:348–361, 1920.

[29] G. Das, K.-I. Lin, H. Mannila, G. Renganathan, and P. Smyth. Rule discovery from times series. In *Proceedings of the Fourth International Conference on Knowledge Discovery and Data Mining (KDD'98)*, pages 16–22, New York, New York, August 1998.

[30] G. Dong and J. Li. Interestingness of discovered association rules in terms of neighborhood-based unexpectedness. In X. Wu, R. Kotagiri, and K. Korb, editors, *Proceedings of the Second Pacific-Asia Conference on Knowledge Discovery and Data Mining (PAKDD'98)*, pages 72–86, Melbourne, Australia, April 1998.

[31] L. Egghe and R. Rousseau. Transfer principles and a classification of concentration measures. *Journal of the American Society for Information Science*, 42(7):479–489, 1991.

[32] M. Ester, H.-P. Kriegel, J. Sander, and X. Xu. A density-based algorithm for discovering clusters in large spatial databases with noise. In *Proceedings of the Second International Conference on Knowledge Discovery and Data Mining (KDD'96)*, pages 226–231, Portland, Oregon, August 1996.

[33] U.M. Fayyad, G. Piatetsky-Shapiro, and P. Smyth. From data mining to knowledge discovery. In U.M. Fayyad, G. Piatetsky-Shapiro, P. Smyth, and R. Uthurusamy, editors, *Advances in Knowledge Discovery and Data Mining*, pages 1–34. AAAI/MIT Press, 1996.

[34] S.E. Fienberg. *The Analysis of Cross-Classified Categorical Data*. MIT Press, 1978.

[35] D.J. Fifield. Distributed tree construction from large datasets. Master's thesis, Australian National University, 1992.

[36] D.H. Fisher. Knowledge acquisition via incremental conceptual clustering. *Machine Learning*, 2:139–172, 1987.

[37] W.J. Frawley, G. Piatetsky-Shapiro, and C.J. Matheus. Knowledge discovery in databases: An overview. In *Knowledge Discovery in Databases*, pages 1–27. AAAI/MIT Press, 1991.

[38] A.A. Freitas. On objective measures of rule surprisingness. In J. Zytkow and M. Quafafou, editors, *Proceedings of the Second European Conference on the Principles of Data Mining and Knowledge Discovery (PKDD'98)*, pages 1–9, Nantes, France, September 1998.

[39] P. Gago and C. Bentos. A metric for selection of the most promising rules. In J. Zytkow and M. Quafafou, editors, *Proceedings of the Second European Conference on the Principles of Data Mining and Knowledge Discovery (PKDD '98)*, pages 19–27, Nantes, France, September 1998.

[40] J. Gaston. *The Reward System in British and American Science.* Wiley and Sons, 1978.

[41] R. Godin, R. Missaoui, and H. Alaoui. Incremental concept formation algorithms based on Galois (concept) lattices. *Computational Intelligence,* 11(2):246–267, 1995.

[42] L.A. Goodman. *The Analysis of Cross-Classified Data Having Ordered Categories.* Harvard University Press, 1984.

[43] L.A. Goodman and W.H. Kruskal. *Measures of Association for Cross Classifications.* Springer-Verlag, 1979.

[44] B. Gray and M.E. Orlowska. CCAIIA: Clustering categorical attributes into interesting association rules. In X. Wu, R. Kotagiri, and K. Korb, editors, *Proceedings of the Second Pacific-Asia Conference on Knowledge Discovery and Data Mining (PAKDD '98)*, pages 132–143, Melbourne, Australia, April 1998.

[45] J. Gray, A. Bosworth, A. Layman, and H. Pirahesh. Data cube: A relational aggregation operator generalizing group-by, cross-tab, and sub-totals. In *Proceedings of the 12th International Conference on Data Engineering*, pages 152–159, New Orleans, Louisiana, February 1996.

[46] J.H. Greenberg. The measurement of linguistic diveristy. *Language*, 32:109–115, 1956.

[47] F.O. Gur-Ali and W.A. Wallace. Are we losing accuracy while gaining confidence in induced rules: An assessment of PrIL. In *Proceedings of the First International Conference on Knowledge Discovery and Data Mining (KDD '95)*, pages 9–14, Montreal, Canada, August 1995.

[48] V. Guralnik, D. Wijesekera, and J. Srivastava. Pattern directed mining of sequence data. In *Proceedings of the Fourth International Conference on Knowledge Discovery and Data Mining (KDD '98)*, pages 51–57, New York, New York, August 1998.

[49] H.J. Hamilton and D.F. Fudger. Estimating DBLearn's potential for knowledge discovery in databases. *Computational Intelligence,* 11(2):280–296, 1995.

[50] H.J. Hamilton, R.J. Hilderman, and N. Cercone. Attribute-oriented induction using domain generalization graphs. In *Proceedings of the Eighth IEEE International Conference on Tools with Artificial Intelligence (ICTAI '96)*, pages 246–253, Toulouse, France, November 1996.

[51] H.J. Hamilton, R.J. Hilderman, L. Li, and D.J. Randall. Generalization lattices. In J. Zytkow and M. Quafafou, editors, *Proceedings of the Second European Conference on the Principles of Data Mining and Knowledge Discovery (PKDD '98)*, pages 328–336, Nantes, France, September 1998.

[52] H.J. Hamilton, N. Shan, and W. Ziarko. Machine learning of credible classifications. In A. Sattar, editor, *Proceedings of the Tenth Australian Conference on Artificial Intelligence (AI '97)*, pages 330–339, Perth, Australia, November/December 1997. Springer Verlag.

[53] E.-H. Han, G. Karypis, and V. Kumar. Scalable parallel data mining for association rules. In *Proceedings of the ACM SIGMOD International Conference on Management of Data (SIGMOD'97)*, pages 277–288, May 1997.

[54] J. Han. Towards efficient induction mechanisms in database systems. *Theoretical Computer Science*, 133:361–385, October 1994.

[55] J. Han, Y. Cai, and N. Cercone. Knowledge discovery in databases: An attribute-oriented approach. In *Proceedings of the 18th International Conference on Very Large Data Bases*, pages 547–559, Vancouver, Canada, August 1992.

[56] J. Han, Y. Cai, and N. Cercone. Data-driven discovery of quantitative rules in relational databases. *IEEE Transactions on Knowledge and Data Engineering*, 5(1):29–40, February 1993.

[57] J. Han and Y. Fu. Discovery of multiple-level association rules from large databases. In *Proc. of 1995 Int'l Conf. on Very Large Data Bases (VLDB'95)*, 1995.

[58] J. Han, Y. Fu, and S. Tang. Advances of the DBLearn system for knowledge discovery in large databases. In *Proc. 1995 Int'l Joint Conf. on Artificial Intelligence (IJCAI'95)*, pages 2049–2050, 1995.

[59] J. Han, W. Ging, and Y. Yin. Mining segment-wise periodic patterns in time-related databases. In *Proceedings of the Fourth International Conference on Knowledge Discovery and Data Mining (KDD'98)*, pages 214–218, New York, New York, August 1998.

[60] J. Han and M. Kamber. *Data Mining: Concepts and Techniques*. Morgan Kaufmann, 2001.

[61] P.E. Hart. Entropy and other measures of concentration. *Journal of the Royal Statistical Society, Series A*, 134:73–85, 1971.

[62] J.L. Hexter and J.W. Snow. An entropy measure of relative aggregate concentration. *Southern Economic Journal*, 36:239–243, 1970.

[63] R.J. Hilderman, C.L. Carter, H.J. Hamilton, and N. Cercone. Mining association rules from market basket data using share measures and characterized itemsets. *International Journal on Artificial Intelligence Tools*, 7(2):189–220, June 1998.

[64] R.J. Hilderman, C.L. Carter, H.J. Hamilton, and N. Cercone. Mining market basket data using share measures and characterized itemsets. In X. Wu, R. Kotagiri, and K. Korb, editors, *Proceedings of the Second Pacific-Asia Conference on Knowledge Discovery and Data Mining (PAKDD'98)*, pages 159–173, Melbourne, Australia, April 1998.

[65] R.J. Hilderman and H.J. Hamilton. Heuristic measures of interestingness. In J. Zytkow and J. Rauch, editors, *Proceedings of the Third European Conference on the Principles of Data Mining and Knowledge Discovery (PKDD'99)*, pages 232–241, Prague, Czech Republic, September 1999.

[66] R.J. Hilderman and H.J. Hamilton. Heuristics for ranking the interestingness of discovered knowledge. In N. Zhong and L. Zhou, editors, *Proceedings of the Third Pacific-Asia Conference on Knowledge Discovery and Data Mining (PAKDD'99)*, pages 204–209, Beijing, China, April 1999.

[67] R.J. Hilderman and H.J. Hamilton. Applying objective interestingness measures in data mining systems. In D.A. Zighed, J. Komorowski, and J. Zytkow, editors, *Proceedings of the Fourth European Symposium on Principles of Data Mining and Knowledge Discovery (PKDD'00)*, pages 432–439, Lyon, France, September 2000.

[68] R.J. Hilderman and H.J. Hamilton. Principles for mining summaries using objective measures of interestingness. In *Proceedings of the Twelfth IEEE International Conference on Tools with Artificial Intelligence (ICTAI'00)*, pages 72–81, Vancouver, Canada, November 2000.

[69] R.J. Hilderman and H.J. Hamilton. Evaluation of interestingness measures for ranking discovered knowledge. In D. Cheung, G.J. Williams, and Q. Li, editors, *Proceedings of the Fifth Pacific-Asia Conference on Knowledge Discovery and Data Mining (PAKDD'01)*, pages 247–259, Hong Kong, April 2001.

[70] R.J. Hilderman, H.J. Hamilton, and B. Barber. Ranking the interestingness of summaries from data mining systems. In *Proceedings of the 12th International Florida Artificial Intelligence Research Symposium (FLAIRS'99)*, pages 100–106, Orlando, Florida, May 1999.

[71] R.J. Hilderman, H.J. Hamilton, and N. Cercone. Data mining in large databases using domain generalization graphs. *Journal of Intelligent Information Systems*, 13(3), November 1999.

[72] R.J. Hilderman, H.J. Hamilton, R.J. Kowalchuk, and N. Cercone. Parallel knowledge discovery using domain generalization graphs. In J. Komorowski and J. Zytkow, editors, *Proceedings of the First European Conference on the Principles of Data Mining and Knowledge Discovery (PKDD'97)*, pages 25–35, Trondheim, Norway, June 1997.

[73] R.J. Hilderman, L. Li, and H.J. Hamilton. Visualizing data mining results with domain generalization graphs. In U. Fayyad, G.G. Grinstein, and A. Wierse, editors, *Information Visualization in Data Mining and Knowledge Discovery*. Morgan Kaufmann Publishers, 2001. In press.

[74] M.O. Hill. Diversity and evenness: A unifying notation and its consequences. *Ecology*, 54:427–432, 1973.

[75] A. Horowitz and I. Horowitz. Entropy Markov processes and competition in the brewing industry. *Journal of Industrial Economics*, 16:196–211, 1968.

[76] S.M. Hurlbert. The nonconcept of species diversity: A critique and alternative parameters. *Ecology*, 52:577–586, 1971.

[77] H.-Y. Hwang and W.-C. Fu. Efficient algorithms for attribute-oriented induction. In *Proceedings of the First International Conference on Knowledge Discovery and Data Mining (KDD'95)*, pages 168–173, Montreal, Canada, August 1995.

[78] J. Iszak. Sensitivity profiles of diveristy indices. *Biometrical Journal*, 38(8):921–930, 1996.

[79] M. Kamber and R. Shinghal. Evaluating the interestingness of characteristic rules. In *Proceedings of the Second International Conference on Knowledge Discovery and Data Mining (KDD'96)*, pages 263–266, Portland, Oregon, August 1996.

[80] L. Kaufman and P.J. Rousseeuw. *Finding Groups in Data: An Introduction to Cluster Analysis.* Wiley and Sons, 1978.

[81] E.J. Keogh and M.J. Pazzani. An enhanced representation of time series which allows fast and accurate classification, clustering, and relevance feedback. In *Proceedings of the Fourth International Conference on Knowledge Discovery and Data Mining (KDD'98)*, pages 239–243, New York, New York, August 1998.

[82] M. Klemettinen, H. Mannila, P. Ronkainen, H. Toivonen, and A.I. Verkamo. Finding interesting rules from large sets of discovered association rules. In N.R. Adam, B.K. Bhargava, and Y. Yesha, editors, *Proceedings of the Third International Conference on Information and Knowledge Management*, pages 401–407, Gaitersburg, Maryland, 1994.

[83] A.J. Knobbe and P.W. Adrians. Analyzing binary associations. In *Proceedings of the Second International Conference on Knowledge Discovery and Data Mining (KDD'96)*, pages 311–314, Portland, Oregon, August 1996.

[84] S. Kullback and R.A. Leibler. On information and sufficiency. *Annals of Mathematical Statistics*, 22:79–86, 1951.

[85] R.C. Lewontin. The apportionment of human diversity. *Evolutionary Biology*, 6:381–398, 1972.

[86] S. Lieberson. An extension of Greenberg's linguistic diversity measures. *Language*, 40:526–531, 1964.

[87] A.M. Liebetrau. *Measures of Association.* Sage Publications, 1983.

[88] B. Liu, W. Hsu, and S. Chen. Using general impressions to analyze discovered classification rules. In *Proceedings of the Third International Conference on Knowledge Discovery and Data Mining (KDD'97)*, pages 31–36, Newport Beach, California, August 1997.

[89] H. Liu, H. Lu, L. Feng, and F. Hussain. Efficient search of reliable exceptions. In N. Zhong and L. Zhou, editors, *Proceedings of the Third Pacific-Asia Conference on Knowledge Discovery and Data Mining (PAKDD'99)*, pages 194–203, Beijing, China, April 1999.

[90] H. Liu, H. Lu, and J. Yao. Identifying relevant databases for multidatabase mining. In X. Wu, R. Kotagiri, and K. Korb, editors, *Proceedings of the Second Pacific-Asia Conference on Knowledge Discovery and Data Mining (PAKDD'98)*, pages 210–221, Melbourne, Australia, April 1998.

[91] R.H. MacArthur. Patterns of species diversity. *Biological Review*, 40:510–533, 1965.

[92] A.E. Magurran. *Ecological Diversity and Its Measurement.* Princeton University Press, 1988.

[93] H. Mannila and H. Toivonen. Discovering generalized episodes using minimal occurrences. In *Proceedings of the Second International Conference on Knowledge Discovery and Data Mining (KDD'96)*, pages 146–151, Portland, Oregon, August 1996.

[94] H. Mannila, H. Toivonen, and A.I. Verkamo. Discovering frequent episodes in sequences. In *Proceedings of the First International Conference on Knowledge Discovery and Data Mining (KDD'95)*, pages 210–215, Montreal, Canada, August 1995.

[95] C.J. Matheus and G. Piatetsky-Shapiro. Selecting and reporting what is interesting: The KEFIR application to healthcare data. In U.M. Fayyad, G. Piatetsky-Shapiro, P. Smyth, and R. Uthurusamy, editors, *Advances in Knowledge Discovery and Data Mining*, pages 401–419, Menlo Park, California, 1996. AAAI Press/MIT Press.

[96] R.P. McIntosh. An index of diversity and the relation of certain concepts to diveristy. *Ecology*, 48(3):392–404, 1967.

[97] M. Mehta, R. Agrawal, and J. Rissanen. SLIQ: A fast scalable classifier for data mining. In *Proceedings of the Fifth International Conference on Extending Database Technology (EDBT'96)*, pages 18–32, Avignon, France, March 1996.

[98] R.S. Michalski, I. Mozetic, J. Hong, and N. Lavrac. The multi-purpose incremental learning system AQ15 and its testing application to three medical domains. In *Proceedings of the Fifth National Conference on Artificial Intelligence*, pages 1041–1045, 1986.

[99] R.S. Michalski and R.E. Stepp. Learning from observation: Conceptual clustering. In R.S. Michalski, J.G. Carbonell, and T.M. Mitchell, editors, *Machine Learning: An Artificial Intelligence Approach*, pages 331–363. Tioga Publishing Company, 1983.

[100] T.M. Mitchell. *Version spaces: An approach to concept learning*. PhD thesis, Stanford University, 1978.

[101] T.M. Mitchell. Generalization as search. *Artificial Intelligence*, 18(2):203–226, 1982.

[102] J. Molinari. A calibrated index for the measurement of evenness. *Oikos*, 56(3):319–326, 1989.

[103] R.T. Ng and J. Han. Efficient and effective clustering methods for spatial data mining. In *Proceedings of the 20th International Conference on Very Large Databases (VLDB'94)*, pages 144–155, Santiago, Chile, September 1994.

[104] B. Padmanabhan and A. Tuzhilin. Pattern discovery in temporal databases: A temporal logic approach. In *Proceedings of the Second International Conference on Knowledge Discovery and Data Mining (KDD'96)*, pages 351–354, Portland, Oregon, August 1996.

[105] B. Padmanabhan and A. Tuzhilin. A belief-driven method for discovering unexpected patterns. In *Proceedings of the Fourth International Conference on Knowledge Discovery and Data Mining (KDD'98)*, pages 94–100, New York, New York, August 1998.

[106] J.S. Park, M.-S. Chen, and P.S. Yu. An effective hash-based algorithm for mining association rules. *SIGMOD Record*, 25(2):175–186, 1995.

[107] D. Partridge and W. Krzanowski. Software diversity: Practical statistics for its measurement and exploitation. *Information and Software Technology*, 39:707–717, 1997.

[108] G.P. Patil and C. Taillie. Diversity as a concept and its measurement. *Journal of the American Statistical Association*, 77(379):548–567, 1982.

[109] R.K. Peet. The measurement of species diversity. *Annual Review of Ecology and Systematics*, 5:285–307, 1974.

[110] G. Piatetsky-Shapiro. Discovery, analysis and presentation of strong rules. In *Knowledge Discovery in Databases*, pages 229–248. AAAI/MIT Press, 1991.

[111] A.D. Pratt. A measure of class concentration in bibliometrics. *Journal of the American Society for Information Science*, 28:285–292, 1977.

[112] J. R. Quinlan. *C4.5 Programs for Machine Learning*. Morgan Kaufmann, 1993.

[113] J.R. Quinlan. Induction of decision trees. *Machine Learning*, 1:81–106, 1986.

[114] D. Rafiei and A. Mendelzon. Similarity-based queries for time series data. In *Proceedings of the ACM SIGMOD International Conference on Management of Data (SIGMOD'97)*, pages 13–23, May 1997.

[115] D.J. Randall, H.J. Hamilton, and R.J. Hilderman. Temporal generalization with domain generalization graphs. *International Journal of Pattern Recognition and Artificial Intelligence*. To appear.

[116] R.B. Rao, S. Rickard, and F. Coetzee. Time series forecasting from high-dimensional data with multiple adaptive layers. In *Proceedings of the Fourth International Conference on Knowledge Discovery and Data Mining (KDD'98)*, pages 319–323, New York, New York, August 1998.

[117] J.L. Ray and J.D. Singer. Measuring the concentration of power in the international system. *Sociological Methods and Research*, 1:403–437, 1973.

[118] H.T. Reynolds. *The Analysis of Cross-Classifications*. Free Press, 1977.

[119] J. Rissanen. *Stochastic Complexity in Statistical Inquiry*. World Scientific Publishing Company, 1989.

[120] W.A. Rosenkrantz. *Introduction to Probability and Statistics for Scientists and Engineers*. McGraw-Hill, 1997.

[121] A.P. Sanjeev and J. Zytkow. Discovering enrollment knowledge in university databases. In *Proceedings of the First International Conference on Knowledge Discovery and Data Mining (KDD'95)*, pages 246–251, Montreal, Canada, August 1995.

[122] A. Savasere, E. Omiecinski, and S. Navathe. An efficient algorithm for mining association rules in large databases. In *Proceedings of the 21th International Conference on Very Large Databases (VLDB'95)*, pages 432–444, Zurich, Switzerland, September 1995.

[123] R.R. Schutz. On the measurement of income inequality. *American Economic Review*, 41:107–122, March 1951.

[124] J. Shafer, R. Agrawal, and M. Mehta. SPRINT: A scalable parallel classifier for data mining. In *Proceedings of the 22nd International Conference on Very Large Databases (VLDB'96)*, pages 544–555, Mumbay, India, September 1996.

[125] C.E. Shannon and W. Weaver. *The Mathematical Theory of Communication*. University of Illinois Press, 1949.

[126] A. Silberschatz and A. Tuzhilin. On subjective measures of interestingness in knowledge discovery. In *Proceedings of the First International Conference on Knowledge Discovery and Data Mining (KDD'95)*, pages 275–281, Montreal, Canada, August 1995.

[127] E.H. Simpson. Measurement of diversity. *Nature*, 163:688, 1949.

[128] P. Smyth and R.M. Goodman. Rule induction using information theory. In *Knowledge Discovery in Databases*, pages 159–176. AAAI/MIT Press, 1991.

[129] R. Srikant and R. Agrawal. Mining generalized association rules. In *Proceedings of the 21th International Conference on Very Large Databases (VLDB'95)*, pages 407–419, Zurich, Switzerland, September 1995.

[130] R. Srikant and R. Agrawal. Mining sequential patterns: Generalization and performance improvements. In *Proceedings of the Fifth International Conference on Extending Database Technology (EDBT'96)*, Avignon, France, March 1996.

[131] R. Srikant, Q. Vu, and R. Agrawal. Mining association rules with item constraints. In *Proceedings of the Third International Conference on Knowledge Discovery and Data Mining (KDD'97)*, pages 67–73, Newport Beach, California, August 1997.

[132] G. Stumme, R. Wille, and U. Wille. Conceptual knowledge discovery in databases using formal concept analysis methods. In J. Zytkow and M. Quafafou, editors, *Proceedings of the Second European Conference on the Principles of Data Mining and Knowledge Discovery (PKDD'98)*, pages 450–458, Nantes, France, September 1998.

[133] H. Theil. *Economics and Information Theory*. Rand McNally, 1970.

[134] H. Toivonen. Sampling large databases for finding association rules. In *Proceedings of the 22nd International Conference on Very Large Databases (VLDB'96)*, pages 134–145, Mumbay, India, September 1996.

[135] W. Wang, J. Yang, and R. Muntz. STING: A statistical information grid approach to spatial data mining. In *Proceedings of the 23nd International Conference on Very Large Databases (VLDB'97)*, pages 186–195, Athens, Greece, September 1997.

[136] G.M. Weiss and H. Hirsh. Learning to predict rare events in event sequences. In *Proceedings of the Fourth International Conference on Knowledge Discovery and Data Mining (KDD'98)*, pages 359–363, New York, New York, August 1998.

[137] M.L. Weitzman. On diversity. *The Quarterly Journal of Economics*, pages 363–405, May 1992.

[138] R.H. Whittaker. Evolution and measurement of species diversity. *Taxon*, 21 (2/3):213–251, May 1972.

[139] J.F. Young. *Information Theory*. John Wiley & Sons, 1971.

[140] M.J. Zaki, N. Lesh, and M. Ogihara. PlanMine: Sequence mining for plan failures. In *Proceedings of the Fourth International Conference on Knowledge Discovery and Data Mining (KDD'98)*, pages 369–373, New York, New York, August 1998.

[141] M.J. Zaki, S. Parthasarathy, M. Ogihara, and W. Li. New algorithms for fast discovery of association rules. In *Proceedings of the Third International Conference on Knowledge*

Discovery and Data Mining (KDD'97), pages 283–286, Newport Beach, California, August 1997.

[142] R. Zembowicz and J. Zytkow. From contingency tables to various forms of knowledge in databases. In U.M. Fayyad, G. Piatetsky-Shapiro, P. Smyth, and R. Uthurusamy, editors, *Advances in Knowledge Discovery and Data Mining*, pages 329–349, Menlo Park, California, 1996. AAAI Press/MIT Press.

[143] J. Zhang and R.S. Michalski. An integration of rule induction and exemplar-based learning for graded concepts. *Machine Learning*, 21:235–267, 1995.

[144] T. Zhang, R. Ramakrishnan, and M. Livny. BIRCH: An efficient data clustering method for very large databases. In *Proceedings of the ACM SIGMOD International Conference on Management of Data (SIGMOD'96)*, pages 103–114, June 1996.

[145] N. Zhong, Y.Y. Yao, and S. Ohsuga. Peculiarity-oriented multi-database mining. In J. Zytkow and J. Rauch, editors, *Proceedings of the Third European Conference on the Principles of Data Mining and Knowledge Discovery (PKDD'99)*, pages 136–146, Prague, Czech Republic, September 1999.

Appendix A
Comparison of Assigned Ranks

In Tables A.1 through A.7, the *Summary ID* column describes a unique summary identifier (for reference purposes), the *Non-ANY Attributes* column describes the number of attributes that have not been generalized to the level of the root node in the associated DGG (which contains the default description "ANY"), the *No. of Tuples* column describes the number of tuples in the summary, and the *Index* and *Rank* columns describe the calculated index value and the assigned rank, respectively, as determined by the corresponding measure.

Table A.1.   Ranks assigned by $I_{Shannon}$ and I_{Total} from *N-2*

Summary ID	Non-ANY Attributes	No. of Tuples	$I_{Shannon}$ Index	Rank	I_{Total} Index	Rank
1	1	2	0.348869	1.5	0.697738	1.5
2	1	3	0.866330	5.0	2.598990	5.0
3	1	4	0.443306	3.5	1.773225	3.5
4	1	5	1.846288	10.0	9.231440	7.0
5	1	6	2.125994	11.0	12.755962	9.0
6	1	9	2.268893	13.0	20.420033	13.0
7	1	10	1.419260	8.5	14.192604	10.5
8	2	2	0.348869	1.5	0.697738	1.5
9	2	4	0.443306	3.5	1.773225	3.5
10	2	5	1.215166	6.0	6.075830	6.0
11	2	9	2.194598	12.0	19.751385	12.0
12	2	9	1.309049	7.0	11.781437	8.0
13	2	10	1.419260	8.5	14.192604	10.5
14	2	11	2.473949	16.0	27.213436	14.0
15	2	16	2.616697	18.0	41.867161	16.0
16	2	17	2.288068	15.0	38.897160	15.0
17	2	21	2.567410	17.0	53.915619	18.0
18	2	21	2.282864	14.0	47.940136	17.0
19	2	30	2.710100	19.0	81.302986	19.0
20	2	40	3.259974	20.0	130.39897	20.0
21	2	50	3.538550	21.0	176.92749	21.0
22	2	67	3.679394	22.0	246.51939	22.0

Table A.2. Ranks assigned by I_{Max} and $I_{McIntosh}$ from N-2

Summary ID	Non-ANY Attributes	No. of Tuples	I_{Max} Index	I_{Max} Rank	$I_{McIntosh}$ Index	$I_{McIntosh}$ Rank
1	1	2	1.000000	1.5	0.063874	1.5
2	1	3	1.584963	3.0	0.233956	5.0
3	1	4	2.000000	4.5	0.065254	3.5
4	1	5	2.321928	6.5	0.458697	10.0
5	1	6	2.584963	8.0	0.496780	14.0
6	1	9	3.169925	10.0	0.501894	15.0
7	1	10	3.321928	12.5	0.314518	8.5
8	2	2	1.000000	1.5	0.063874	1.5
9	2	4	2.000000	4.5	0.065254	3.5
10	2	5	2.321928	6.5	0.282728	6.0
11	2	9	3.169925	10.0	0.494505	12.0
12	2	9	3.169925	10.0	0.283782	7.0
13	2	10	3.321928	12.5	0.314518	8.5
14	2	11	3.459432	14.0	0.529937	16.0
15	2	16	4.000000	15.0	0.534837	18.0
16	2	17	4.087463	16.0	0.495313	13.0
17	2	21	4.392317	17.5	0.530693	17.0
18	2	21	4.392317	17.5	0.477246	11.0
19	2	30	4.906891	19.0	0.535592	19.0
20	2	40	5.321928	20.0	0.630569	20.0
21	2	50	5.643856	21.0	0.657900	21.0
22	2	67	6.066089	22.0	0.661515	22.0

Table A.3.   Ranks assigned by I_{Lorenz} and I_{Berger} from *N-2*

Summary ID	Non-ANY Attributes	No. of Tuples	I_{Lorenz} Index	Rank	I_{Berger} Index	Rank
1	1	2	0.532746	1.5	0.934509	2.5
2	1	3	0.429060	3.0	0.712931	5.0
3	1	4	0.277279	7.5	0.934509	2.5
4	1	5	0.402945	4.0	0.393841	12.0
5	1	6	0.379616	5.0	0.393841	12.0
6	1	9	0.261123	9.0	0.393841	12.0
7	1	10	0.165982	14.5	0.603704	8.5
8	2	2	0.532746	1.5	0.934509	2.5
9	2	4	0.277279	7.5	0.934509	2.5
10	2	5	0.283677	6.0	0.666853	6.5
11	2	9	0.253015	10.0	0.365614	16.5
12	2	9	0.166537	13.0	0.666853	6.5
13	2	10	0.165982	14.5	0.603704	8.5
14	2	11	0.236883	11.0	0.365614	16.5
15	2	16	0.175297	12.0	0.365614	16.5
16	2	17	0.142521	16.0	0.365614	16.5
17	2	21	0.132651	17.0	0.365614	16.5
18	2	21	0.118036	18.0	0.420841	10.0
19	2	30	0.100625	21.0	0.365614	16.5
20	2	40	0.108058	19.0	0.234297	21.0
21	2	50	0.102211	20.0	0.234297	21.0
22	2	67	0.083496	22.0	0.234297	21.0

Table A.4. Ranks assigned by I_{Schutz} and I_{Bray} from N-2

Summary ID	Non-ANY Attributes	No. of Tuples	I_{Schutz}		I_{Bray}	
			Index	Rank	Index	Rank
1	1	2	0.434509	4.5	0.565491	4.5
2	1	3	0.379598	3.0	0.620402	3.0
3	1	4	0.684509	11.5	0.315491	11.5
4	1	5	0.310744	2.0	0.689256	2.0
5	1	6	0.294042	1.0	0.705958	1.0
6	1	9	0.466300	6.0	0.533700	6.0
7	1	10	0.734509	19.5	0.265491	19.5
8	2	2	0.434509	4.5	0.565491	4.5
9	2	4	0.684509	11.5	0.315491	11.5
10	2	5	0.534397	9.0	0.465603	9.0
11	2	9	0.516940	8.0	0.483060	8.0
12	2	9	0.712175	15.0	0.287825	15.0
13	2	10	0.734509	19.5	0.265491	19.5
14	2	11	0.486637	7.0	0.513363	7.0
15	2	16	0.600273	10.0	0.399727	10.0
16	2	17	0.699103	14.0	0.300897	14.0
17	2	21	0.696302	13.0	0.303698	13.0
18	2	21	0.743921	22.0	0.256079	22.0
19	2	30	0.723102	16.0	0.276898	16.0
20	2	40	0.734397	17.5	0.265603	17.5
21	2	50	0.734397	17.5	0.265603	17.5
22	2	67	0.742610	21.0	0.25739	21.0

Table A.5.   Ranks assigned by $I_{Whittaker}$ and $I_{Kullback}$ from N-2

Summary ID	Non-ANY Attributes	No. of Tuples	$I_{Whittaker}$ Index	$I_{Whittaker}$ Rank	$I_{Kullback}$ Index	$I_{Kullback}$ Rank
1	1	2	0.565491	4.5	0.348869	1.5
2	1	3	0.620402	3.0	0.866330	5.0
3	1	4	0.315491	11.5	0.443306	3.5
4	1	5	0.689256	2.0	1.846288	10.0
5	1	6	0.705958	1.0	2.125994	11.0
6	1	9	0.533700	6.0	2.268893	13.0
7	1	10	0.265491	19.5	1.419260	8.5
8	2	2	0.565491	4.5	0.348869	1.5
9	2	4	0.315491	11.5	0.443306	3.5
10	2	5	0.465603	9.0	1.215166	6.0
11	2	9	0.483060	8.0	2.194598	12.0
12	2	9	0.287825	15.0	1.309049	7.0
13	2	10	0.265491	19.5	1.419260	8.5
14	2	11	0.513363	7.0	2.473949	16.0
15	2	16	0.399727	10.0	2.616697	18.0
16	2	17	0.300897	14.0	2.288068	15.0
17	2	21	0.303698	13.0	2.567410	17.0
18	2	21	0.256079	22.0	2.282864	14.0
19	2	30	0.276898	16.0	2.710100	19.0
20	2	40	0.265603	17.5	3.259974	20.0
21	2	50	0.265603	17.5	3.538550	21.0
22	2	67	0.257390	21.0	3.679394	22.0

Table A.6.   Ranks assigned by $I_{MacArthur}$ and I_{Theil} from N-2

Summary ID	Non-ANY Attributes	No. of Tuples	$I_{MacArthur}$ Index	$I_{MacArthur}$ Rank	I_{Theil} Index	I_{Theil} Rank
1	1	2	0.184731	3.5	0.651131	1.5
2	1	3	0.218074	5.0	0.718633	3.0
3	1	4	0.399511	11.5	1.556694	7.5
4	1	5	0.144729	2.0	0.757153	4.0
5	1	6	0.132377	1.0	0.777902	5.0
6	1	9	0.243857	6.0	1.710559	9.0
7	1	10	0.457814	16.5	2.508888	13.5
8	2	2	0.184731	3.5	0.651131	1.5
9	2	4	0.399511	11.5	1.556694	7.5
10	2	5	0.298402	9.0	1.195810	6.0
11	2	9	0.264620	8.0	1.898130	10.0
12	2	9	0.452998	15.0	2.249471	12.0
13	2	10	0.457814	16.5	2.508888	13.5
14	2	11	0.260255	7.0	2.025527	11.0
15	2	16	0.342143	10.0	2.939297	15.0
16	2	17	0.441534	14.0	3.512838	16.0
17	2	21	0.440642	13.0	3.890191	17.0
18	2	21	0.487441	20.0	3.982314	18.0
19	2	30	0.494412	21.0	4.485426	19.0
20	2	40	0.479347	18.0	5.317662	20.0
21	2	50	0.482560	19.0	5.751495	21.0
22	2	67	0.515363	22.0	6.181546	22.0

Table A.7.   Ranks assigned by $I_{Atkinson}$ and I_{Gini} from *N-2*

Summary ID	Non-ANY Attributes	No. of Tuples	$I_{Atkinson}$		I_{Gini}	
			Index	Rank	Index	Rank
1	1	2	0.505218	1.5	0.266373	1.5
2	1	3	0.914901	22.0	0.214530	5.0
3	1	4	0.792127	8.5	0.138640	3.5
4	1	5	0.759314	6.0	0.201473	8.0
5	1	6	0.693136	3.0	0.189808	11.0
6	1	9	0.765973	7.0	0.130562	13.0
7	1	10	0.821439	11.5	0.082991	9.5
8	2	2	0.505218	1.5	0.266373	1.5
9	2	4	0.792127	8.5	0.138640	3.5
10	2	5	0.859044	16.0	0.141839	6.0
11	2	9	0.759162	5.0	0.126508	12.0
12	2	9	0.884562	19.0	0.083269	7.0
13	2	10	0.821439	11.5	0.082991	9.5
14	2	11	0.727091	4.0	0.118442	14.0
15	2	16	0.797472	10.0	0.087649	16.0
16	2	17	0.860465	17.0	0.071261	15.0
17	2	21	0.852812	13.0	0.066326	18.0
18	2	21	0.862917	18.0	0.059018	17.0
19	2	30	0.894697	21.0	0.050313	19.0
20	2	40	0.854864	15.0	0.054029	20.0
21	2	50	0.854329	14.0	0.051106	21.0
22	2	67	0.885877	20.0	0.041730	22.0

Appendix B
Ranking Similarities

In Table B.1, the *Measure 1* and *Measure 2* columns describe the pairs of measures being compared, the *C-2*, *C-3*, *C-4*, and *C-5* columns describe the coefficients corresponding to the pairs of measures from the two-, three-, four-, and five-attribute Customer discovery tasks, respectively, and the *Average* column describes the average of the coefficients for the respective group of discovery tasks.

Table B.1. Ranking similarities for Customer discovery tasks

| Measure 1 | Measure 2 | Gamma Correlation Coefficient | | | | |
		C-2	C-3	C-4	C-5	Average
$I_{Variance}$	$I_{Simpson}$	0.988914	0.990148	0.989673	0.979941	0.987169
$I_{Variance}$	$I_{Shannon}$	0.975610	0.978325	0.973591	0.977704	0.976308
$I_{Variance}$	I_{Total}	0.955654	0.958621	0.947181	0.938105	0.949890
$I_{Variance}$	I_{Max}	0.950617	0.954138	0.941187	0.932472	0.944604
$I_{Variance}$	$I_{McIntosh}$	0.988914	0.990148	0.989673	0.979953	0.987172
$I_{Variance}$	I_{Lorenz}	0.700665	0.635961	0.643812	0.714677	0.673779
$I_{Variance}$	I_{Berger}	0.930036	0.958333	0.940855	0.916278	0.936376
$I_{Variance}$	I_{Schutz}	0.496674	0.277833	0.353310	0.589453	0.429318
$I_{Variance}$	I_{Bray}	0.496674	0.277833	0.353310	0.589453	0.429318
$I_{Variance}$	$I_{Whittaker}$	0.496674	0.277833	0.353310	0.589453	0.429318
$I_{Variance}$	$I_{Kullback}$	0.975610	0.978325	0.973591	0.977704	0.976308
$I_{Variance}$	$I_{MacArthur}$	0.547672	0.363547	0.413239	0.463463	0.446980
$I_{Variance}$	I_{Theil}	0.866962	0.872414	0.854749	0.857911	0.863009
$I_{Variance}$	$I_{Atkinson}$	0.478936	0.322660	0.344168	0.458899	0.401166
$I_{Variance}$	I_{Gini}	0.973392	0.981773	0.978669	0.971184	0.976225
$I_{Simpson}$	$I_{Shannon}$	0.980066	0.986214	0.980203	0.968828	0.978828
$I_{Simpson}$	I_{Total}	0.962348	0.963565	0.949746	0.926630	0.950572
$I_{Simpson}$	I_{Max}	0.957399	0.959143	0.943788	0.920810	0.945285
$I_{Simpson}$	$I_{McIntosh}$	1.000000	1.000000	1.000000	1.000000	1.000000
$I_{Simpson}$	I_{Lorenz}	0.707641	0.641064	0.646531	0.701928	0.674291
$I_{Simpson}$	I_{Berger}	0.925301	0.959423	0.940517	0.907533	0.933194
$I_{Simpson}$	I_{Schutz}	0.503876	0.283112	0.356176	0.577514	0.430170
$I_{Simpson}$	I_{Bray}	0.503876	0.283112	0.356176	0.577514	0.430170
$I_{Simpson}$	$I_{Whittaker}$	0.503876	0.283112	0.356176	0.577514	0.430170
$I_{Simpson}$	$I_{Kullback}$	0.980066	0.986214	0.980203	0.968828	0.978828
$I_{Simpson}$	$I_{MacArthur}$	0.554817	0.368784	0.416075	0.550944	0.472828
$I_{Simpson}$	I_{Theil}	0.873754	0.877400	0.857360	0.845086	0.863400
$I_{Simpson}$	$I_{Atkinson}$	0.486157	0.327917	0.347039	0.447116	0.402057
$I_{Simpson}$	I_{Gini}	0.968992	0.974889	0.972081	0.953126	0.967272
$I_{Shannon}$	I_{Total}	0.968992	0.971443	0.960068	0.954006	0.963627
$I_{Shannon}$	I_{Max}	0.961883	0.965122	0.952699	0.947440	0.956786
$I_{Shannon}$	$I_{McIntosh}$	0.980066	0.986214	0.980203	0.968850	0.978833
$I_{Shannon}$	I_{Lorenz}	0.714286	0.648941	0.655838	0.727214	0.686570
$I_{Shannon}$	I_{Berger}	0.908434	0.949813	0.931252	0.912190	0.719399
$I_{Shannon}$	I_{Schutz}	0.506091	0.289020	0.362098	0.600845	0.439513
$I_{Shannon}$	I_{Bray}	0.506091	0.289020	0.362098	0.600845	0.439513
$I_{Shannon}$	$I_{Whittaker}$	0.506091	0.289020	0.362098	0.600845	0.439513
$I_{Shannon}$	$I_{Kullback}$	1.000000	1.000000	1.000000	1.000000	1.000000
$I_{Shannon}$	$I_{MacArthur}$	0.552602	0.368784	0.418274	0.574391	0.478513
$I_{Shannon}$	I_{Theil}	0.875969	0.883309	0.863283	0.868857	0.872855

Table B.1.   Ranking similarities for Customer discovery tasks (continued)

		Gamma Correlation Coefficient				
Measure 1	Measure 2	C-2	C-3	C-4	C-5	Average
$I_{Shannon}$	$I_{Atkinson}$	0.488372	0.327917	0.349239	0.467777	0.408326
$I_{Shannon}$	I_{Gini}	0.980066	0.978828	0.980711	0.978334	0.979485
I_{Total}	I_{Max}	0.993274	0.994021	0.993145	0.993998	0.993610
I_{Total}	$I_{McIntosh}$	0.962348	0.963565	0.949746	0.926681	0.950585
I_{Total}	I_{Lorenz}	0.745293	0.677499	0.695770	0.773208	0.722943
I_{Total}	I_{Berger}	0.920482	0.947144	0.933846	0.909005	0.927619
I_{Total}	I_{Schutz}	0.537099	0.317578	0.402031	0.645984	0.475673
I_{Total}	I_{Bray}	0.537099	0.317578	0.402031	0.645984	0.475673
I_{Total}	$I_{Whittaker}$	0.537099	0.317578	0.402031	0.645984	0.475673
I_{Total}	$I_{Kullback}$	0.968992	0.971443	0.960068	0.954006	0.963627
I_{Total}	$I_{MacArthur}$	0.583610	0.397341	0.458206	0.620385	0.514886
I_{Total}	I_{Theil}	0.906977	0.911866	0.903215	0.914653	0.710876
I_{Total}	$I_{Atkinson}$	0.519380	0.356475	0.389171	0.512844	0.444468
I_{Total}	I_{Gini}	0.966777	0.962088	0.957360	0.962813	0.962260
I_{Max}	$I_{McIntosh}$	0.957399	0.959143	0.943788	0.920857	0.945297
I_{Max}	I_{Lorenz}	0.771300	0.701545	0.722879	0.800305	0.749007
I_{Max}	I_{Berger}	0.912302	0.940113	0.927882	0.911486	0.922946
I_{Max}	I_{Schutz}	0.558296	0.336323	0.424336	0.668829	0.496971
I_{Max}	I_{Bray}	0.558296	0.336323	0.424336	0.668829	0.496971
I_{Max}	$I_{Whittaker}$	0.558296	0.336323	0.424336	0.668829	0.496971
I_{Max}	$I_{Kullback}$	0.961883	0.965122	0.952699	0.947440	0.956786
I_{Max}	$I_{MacArthur}$	0.609866	0.420030	0.483805	0.646011	0.539928
I_{Max}	I_{Theil}	0.932735	0.937718	0.931962	0.942494	0.936227
I_{Max}	$I_{Atkinson}$	0.544843	0.378675	0.413882	0.536130	0.468383
I_{Max}	I_{Gini}	0.961883	0.957648	0.951500	0.957147	0.957045
$I_{McIntosh}$	I_{Lorenz}	0.707641	0.641064	0.646531	0.701937	0.674293
$I_{McIntosh}$	I_{Berger}	0.925301	0.959423	0.940517	0.907533	0.933194
$I_{McIntosh}$	I_{Schutz}	0.503876	0.283112	0.356176	0.577767	0.429033
$I_{McIntosh}$	I_{Bray}	0.503876	0.283112	0.356176	0.577767	0.429033
$I_{McIntosh}$	$I_{Whittaker}$	0.503876	0.283112	0.356176	0.577767	0.429033
$I_{McIntosh}$	$I_{Kullback}$	0.980066	0.986214	0.980203	0.968850	0.978833
$I_{McIntosh}$	$I_{MacArthur}$	0.554817	0.368784	0.416075	0.551108	0.472696
$I_{McIntosh}$	I_{Theil}	0.873754	0.877400	0.857360	0.844995	0.863377
$I_{McIntosh}$	$I_{Atkinson}$	0.486157	0.327917	0.347039	0.447302	0.402104
$I_{McIntosh}$	I_{Gini}	0.968992	0.974889	0.972081	0.953154	0.967279
I_{Lorenz}	I_{Berger}	0.674699	0.617726	0.641434	0.722523	0.664010
I_{Lorenz}	I_{Schutz}	0.774086	0.633186	0.694416	0.828313	0.732500
I_{Lorenz}	I_{Bray}	0.774086	0.633186	0.694416	0.828313	0.732500
I_{Lorenz}	$I_{Whittaker}$	0.774086	0.633186	0.694416	0.828313	0.732500
I_{Lorenz}	$I_{Kullback}$	0.714286	0.648941	0.655838	0.727214	0.686570

Table B.1. Ranking similarities for Customer discovery tasks (continued)

Measure 1	Measure 2	Gamma Correlation Coefficient				
		C-2	*C-3*	*C-4*	*C-5*	*Average*
I_{Lorenz}	$I_{MacArthur}$	0.829457	0.710980	0.753976	0.840121	0.783634
I_{Lorenz}	I_{Theil}	0.829457	0.756770	0.785787	0.854680	0.806674
I_{Lorenz}	$I_{Atkinson}$	0.760797	0.654357	0.663283	0.717840	0.699070
I_{Lorenz}	I_{Gini}	0.712071	0.639586	0.653130	0.736496	0.685321
I_{Berger}	I_{Schutz}	0.465060	0.237587	0.329566	0.576659	0.402218
I_{Berger}	I_{Bray}	0.465060	0.237587	0.329566	0.576659	0.402218
I_{Berger}	$I_{Whittaker}$	0.465060	0.237587	0.329566	0.576659	0.402218
I_{Berger}	$I_{Kullback}$	0.908434	0.949813	0.931252	0.912190	0.925422
I_{Berger}	$I_{MacArthur}$	0.513253	0.326215	0.396090	0.573715	0.452318
I_{Berger}	I_{Theil}	0.850602	0.869728	0.862504	0.864215	0.861762
I_{Berger}	$I_{Atkinson}$	0.457831	0.295248	0.331048	0.469212	0.388335
I_{Berger}	I_{Gini}	0.908434	0.948745	0.930881	0.911315	0.924844
I_{Schutz}	I_{Bray}	1.000000	1.000000	1.000000	1.000000	1.000000
I_{Schutz}	$I_{Whittaker}$	1.000000	1.000000	1.000000	1.000000	1.000000
I_{Schutz}	$I_{Kullback}$	0.506091	0.289020	0.362098	0.600845	0.439514
I_{Schutz}	$I_{MacArthur}$	0.882614	0.868045	0.875804	0.878879	0.876336
I_{Schutz}	I_{Theil}	0.630122	0.403742	0.498139	0.728885	0.565222
I_{Schutz}	$I_{Atkinson}$	0.800664	0.746430	0.766836	0.767808	0.770435
I_{Schutz}	I_{Gini}	0.503876	0.279665	0.359391	0.609449	0.438095
I_{Bray}	$I_{Whittaker}$	1.000000	1.000000	1.000000	1.000000	1.000000
I_{Bray}	$I_{Kullback}$	0.506091	0.289020	0.362098	0.600845	0.439514
I_{Bray}	$I_{MacArthur}$	0.882614	0.868045	0.875804	0.878889	0.876388
I_{Bray}	I_{Theil}	0.630122	0.403742	0.498139	0.728885	0.565222
I_{Bray}	$I_{Atkinson}$	0.800664	0.746430	0.766836	0.767808	0.770435
I_{Bray}	I_{Gini}	0.503876	0.279665	0.359391	0.609449	0.438095
$I_{Whittaker}$	$I_{Kullback}$	0.506091	0.289020	0.362098	0.600845	0.439514
$I_{Whittaker}$	$I_{MacArthur}$	0.882614	0.868045	0.875804	0.878889	0.876388
$I_{Whittaker}$	I_{Theil}	0.630122	0.403742	0.498139	0.728885	0.565222
$I_{Whittaker}$	$I_{Atkinson}$	0.800664	0.746430	0.766836	0.767808	0.770435
$I_{Whittaker}$	I_{Gini}	0.503876	0.279665	0.359391	0.609449	0.438095
$I_{Kullback}$	$I_{MacArthur}$	0.552602	0.368784	0.418274	0.574391	0.478513
$I_{Kullback}$	I_{Theil}	0.875969	0.883309	0.863283	0.868857	0.872855
$I_{Kullback}$	$I_{Atkinson}$	0.488372	0.327917	0.349239	0.467777	0.408326
$I_{Kullback}$	I_{Gini}	0.980066	0.978828	0.980711	0.978334	0.979485
$I_{MacArthur}$	I_{Theil}	0.676633	0.484490	0.552284	0.703679	0.604272
$I_{MacArthur}$	$I_{Atkinson}$	0.869324	0.844904	0.846362	0.856651	0.854310
$I_{MacArthur}$	I_{Gini}	0.554817	0.363368	0.418613	0.584917	0.480429
I_{Theil}	$I_{Atkinson}$	0.603544	0.437715	0.473773	0.590539	0.526393
I_{Theil}	I_{Gini}	0.873754	0.873954	0.860575	0.878485	0.871692
$I_{Atkinson}$	I_{Gini}	0.490587	0.324471	0.351607	0.478491	0.411289

Appendix C
Summary Complexity

In Tables C.1 and C.2, three rows (labelled T, NA, and CI) are used to describe the relative interestingness of summaries ranked by each measure. The first row describes the average number of tuples contained in each group of summaries, when grouped according to a three-tier scale of relative interestingness (i.e., H=High, M=Moderate, L=Low) by discovery task. The second and third rows describe the corresponding average number of non-ANY attributes in the summaries of each group and the complexity for each group, respectively.

Table C.1.   Relative interestingness versus complexity for C-2 and C-3

Measure		Relative Interestingness					
		C-2			C-3		
		H	M	L	H	M	L
$I_{Variance}$	T	6.8	34.3	70.6	5.0	30.4	70.3
	NA	1.4	1.8	1.9	1.5	1.9	2.4
	CI	9.5	61.7	134.1	7.5	57.8	168.7
$I_{Simpson}$	T	6.8	34.3	70.6	5.0	30.4	70.3
	NA	1.4	1.8	1.9	1.5	2.0	2.4
	CI	9.5	61.7	134.1	7.5	60.8	168.7
$I_{Shannon}$	T	6.8	34.3	70.6	5.0	30.3	70.3
	NA	1.4	1.8	1.9	1.5	2.1	2.4
	CI	9.5	61.7	134.1	7.5	63.6	168.7
I_{Total}	T	6.8	34.3	70.6	5.0	30.5	70.6
	NA	1.4	1.8	1.9	1.5	2.1	2.4
	CI	9.5	61.7	134.1	7.5	64.1	169.4
I_{Max}	T	6.8	34.2	70.6	5.0	30.6	70.6
	NA	1.3	1.8	1.9	1.5	2.2	2.4
	CI	8.8	61.6	134.1	7.5	67.3	169.4
$I_{McIntosh}$	T	6.8	34.3	70.6	5.0	30.4	70.3
	NA	1.4	1.8	1.9	1.5	2.0	2.4
	CI	9.5	61.7	134.1	7.5	60.8	168.7
I_{Lorenz}	T	9.9	33.2	64.9	8.8	26.8	65.1
	NA	1.2	2.0	1.9	1.5	2.3	2.4
	CI	11.9	66.4	123.3	13.2	61.6	156.2
I_{Berger}	T	6.8	33.1	70.1	5.0	31.1	68.2
	NA	1.4	2.0	2.0	1.5	1.8	2.7
	CI	9.5	66.2	140.2	7.5	56.0	184.1
I_{Schutz}	T	16.8	38.3	60.3	17.8	35.6	50.0
	NA	1.3	1.9	1.9	1.7	2.4	2.1
	CI	21.8	72.8	114.6	30.3	85.4	105.0
I_{Bray}	T	16.8	38.3	60.3	17.8	35.6	50.0
	NA	1.3	1.9	1.9	1.7	2.4	2.1
	CI	21.8	72.8	114.6	30.3	85.4	105.0
$I_{Whittaker}$	T	16.8	38.3	60.3	17.8	35.6	50.0
	NA	1.3	1.9	1.9	1.7	2.4	2.1
	CI	21.8	72.8	114.6	30.3	85.4	105.0
$I_{Kullback}$	T	6.8	34.3	70.6	5.0	30.3	70.3
	NA	1.4	1.8	1.9	1.5	2.1	2.4
	CI	9.8	61.0	133.3	7.5	63.9	168.0

Table C.1. Relative interestingness versus complexity for *C-2* and *C-3* (continued)

Measure		Relative Interestingness					
		C-2			C-3		
		H	M	L	H	M	L
$I_{MacArthur}$	T	14.4	30.2	59.9	12.7	28.3	47.7
	NA	1.2	1.8	1.9	1.5	2.3	2.2
	CI	17.3	54.4	113.8	19.1	65.1	104.9
I_{Theil}	T	8.7	34.3	70.4	5.4	30.3	70.5
	NA	1.3	1.9	2.0	1.5	2.2	2.5
	CI	11.3	65.2	140.8	8.1	66.7	176.3
$I_{Atkinson}$	T	14.9	40.0	48.4	16.9	39.1	41.4
	NA	1.1	1.8	1.9	1.3	2.3	2.2
	CI	16.4	72.0	92.0	21.9	89.9	91.1
I_{Gini}	T	6.8	34.3	70.6	5.0	30.3	70.3
	NA	1.4	1.8	1.9	1.5	2.1	2.4
	CI	9.5	61.7	134.1	7.5	63.6	168.7

Table C.2.   Relative interestingness versus complexity for *C-4* and *C-5*

Measure		Relative Interestingness					
		C-4			C-5		
		H	M	L	H	M	L
$I_{Variance}$	T	7.2	33.5	68.9	14.0	22.0	81.2
	NA	2.1	1.0	3.0	2.8	3.5	3.7
	CI	15.1	33.5	206.7	39.2	77.0	300.4
$I_{Simpson}$	T	7.2	33.5	68.9	14.0	22.1	80.1
	NA	2.1	1.0	3.0	2.8	3.5	3.7
	CI	15.1	33.5	206.7	39.2	77.4	296.4
$I_{Shannon}$	T	7.2	33.1	68.9	14.0	22.0	81.2
	NA	2.1	2.0	3.0	2.8	3.5	3.8
	CI	15.1	66.2	206.7	39.2	77.0	308.6
I_{Total}	T	6.9	33.0	69.1	13.6	22.0	81.2
	NA	2.0	2.0	3.0	2.7	3.4	3.8
	CI	13.8	66.0	207.3	32.6	74.8	308.6
I_{Max}	T	6.9	33.0	69.1	13.6	22.0	81.2
	NA	2.0	2.0	3.0	2.7	3.5	3.8
	CI	13.8	66.0	207.3	36.7	77.0	308.6
$I_{McIntosh}$	T	7.2	33.5	68.9	14.0	22.1	80.1
	NA	2.1	1.0	3.0	2.8	3.5	3.7
	CI	15.1	33.5	206.7	39.2	77.4	296.4
I_{Lorenz}	T	11.0	28.9	65.7	17.5	22.1	80.1
	NA	1.9	3.0	2.9	2.6	3.5	3.7
	CI	20.9	86.7	190.5	45.5	77.4	296.8
I_{Berger}	T	7.2	33.3	67.2	14.1	21.6	78.0
	NA	2.1	2.0	3.4	2.7	3.5	4.1
	CI	15.1	66.6	228.5	38.1	75.6	319.8
I_{Schutz}	T	20.1	36.7	49.9	21.8	21.8	77.1
	NA	2.0	3.0	2.6	2.7	3.5	3.6
	CI	40.2	110.1	129.7	58.9	76.3	277.6
I_{Bray}	T	20.1	36.7	49.9	21.8	21.8	77.1
	NA	2.0	3.0	2.6	2.7	3.5	3.6
	CI	40.2	110.1	129.7	58.9	76.3	277.6
$I_{Whittaker}$	T	20.1	36.7	49.9	21.8	21.8	77.1
	NA	2.0	3.0	2.6	2.7	3.5	3.6
	CI	40.2	110.1	129.7	58.9	76.3	277.6
$I_{Kullback}$	T	7.2	33.1	68.9	14.0	22.0	81.2
	NA	2.1	2.0	3.0	2.8	3.5	3.8
	CI	15.0	66.2	204.6	39.1	76.4	305.3
$I_{MacArthur}$	T	15.4	29.3	49.8	20.6	23.5	75.6
	NA	1.8	2.0	2.6	2.6	3.4	3.6
	CI	27.7	58.6	129.5	53.6	79.9	272.2

Table C.2.   Relative interestingness versus complexity for *C-4* and *C-5* (continued)

Measure		Relative Interestingness					
		C-4			C-5		
		H	M	L	H	M	L
I_{Theil}	T	7.9	32.1	68.9	13.8	22.1	81.2
	NA	1.9	3.0	3.0	2.6	3.6	3.8
	CI	15.0	96.3	206.7	35.9	79.6	308.6
$I_{Atkinson}$	T	19.0	39.5	46.4	21.6	22.1	71.9
	NA	1.7	3.0	2.8	2.5	3.5	3.6
	CI	32.3	118.5	129.9	54.0	77.4	258.8
I_{Gini}	T	7.2	33.4	68.9	14.0	21.9	81.2
	NA	2.1	2.8	3.0	2.8	3.5	3.8
	CI	15.1	93.5	206.7	39.2	76.7	305.3

Index

49er 14
All_Gen 26-31, 99, 124, 127
AOG 3, 33
Apriori 12, 13
AQ15 11
attribute-oriented generalization *see AOG*

BIRCH 14

C4.5 2, 12
C5.0 12
CLARANS 14
classification 11, 12, 15, 16, 18-20
CLIQUE 14
CLOUDS 12
CLUSTER/2 13
CN2 11, 12
COBWEB 13
conceptual graphs 9
correlation 14
credibility 19

DBLearn 2
DBSCAN 14
DGG-Discover 25, 99
DGG-Interest 25, 99, 104
DGG 3-7, 9, 25-28, 30-35, 100, 103, 105, 123, 124
DHP 13
DIC 13
distance metric 20, 22
diversity 8, 9, 37-39, 42, 48, 99, 104, 123-125
domain generalization graph *see DGG*

FCLS 12
formal concept analysis 9

Galois lattice 9
general impression 20

hybrid distribution 13

ID3 11, 12
I-measures 17
$I_{Atkinson}$ 46, 108, 114, 121
I_{Berger} 43, 44, 107, 108, 114, 125
I_{Bray} 44, 107, 108, 114, 121, 125
I_{Gini} 43, 107, 108, 114, 121, 125
$I_{Kullback}$ 45, 107, 108, 114, 124, 125
I_{Lorenz} 42, 108, 114, 121, 125
$I_{MacArthur}$ 45, 108, 114, 125
I_{Max} 41, 107, 108, 114, 119, 124
$I_{McIntosh}$ 41, 42, 107, 108, 114, 124, 125
Interestingness 18, 21, 22
I_{Schutz} 44, 107, 108, 114, 117, 119, 125
$I_{Shannon}$ 40, 41, 45, 50, 107, 108, 114, 124, 125
$I_{Simpson}$ 40, 107, 108, 114, 124, 125
Itemset Clustering 13
itemset measures 16
I_{Theil} 45, 46, 107, 108, 114, 124
I_{Total} 41, 107, 108, 114, 124, 125
$I_{Variance}$ 39, 40, 50, 107, 108, 114, 117, 118, 124, 125
$I_{Whittaker}$ 44, 45, 107, 108, 114, 121, 125

J-measure 16

KDD 1, 8, 11, 15, 25, 33
KID3 2, 12
knowledge discovery in databases *see KDD*

Maximum Value Principle 47, 63
Minimum Value Principle 47, 51

Par_All_Gen 30-33
partition 13
peculiarity 23
Permutation Invariance Principle 48, 84

PrIL 12
projected savings 17

Q2 13

reliable exception 22
rule-interest function 15
rule template 16, 17

sampling 13
sequences 15

share 13
SLIQ 12
SPRINT 12
STING 14
Skewness Principle 48, 79
surprisingness 21

time series 15
Transfer Principle 48, 84

VSA 11